THE BIOLOGIST'S HANDBOOK
OF PRONUNCIATIONS

By the same author

A SOURCE-BOOK OF MEDICAL TERMS
  Charles C Thomas, Publisher, Springfield, Illinois

A SOURCE-BOOK OF BIOLOGICAL NAMES AND
TERMS
  Charles C Thomas, Publisher, Springfield, Illinois

THE
BIOLOGIST'S HANDBOOK
OF PRONUNCIATIONS

By
EDMUND C. JAEGER, D.Sc.

Curator of Plants
Riverside (California) Municipal Museum

Illustrations by
Morris Van Dame and The Author

CHARLES C THOMAS • PUBLISHER

Springfield · *Illinois* · *U.S.A*

CHARLES C THOMAS • PUBLISHER

BANNERSTONE HOUSE

301-327 East Lawrence Avenue, Springfield, Illinois, U.S.A.

Published simultaneously in the
British Commonwealth of Nations by

BLACKWELL SCIENTIFIC PUBLICATIONS, LTD.,

OXFORD, ENGLAND

Published simultaneously in Canada by

THE RYERSON PRESS, TORONTO

With THOMAS BOOKS careful attention is given to all details of manufacturing and design. It is the Publisher's desire to present books that are satisfactory as to their physical qualities and artistic possibilities and appropriate for their particular use. THOMAS BOOKS will be true to those laws of quality that assure a good name and good will.

Printed in the United States of America

PREFACE

IT IS A COMPLAINT all too common among those who are beginning the study of the biological sciences that they can make little satisfactory progress in the pronunciation of the numerous scientific names and terms they must use. Their path is made thorny by the fact that advanced students and even many teachers of scientific studies, to whom they look for guidance, mispronounce, often atrociously, many of the terms. Perhaps all, both students and teachers, would gladly improve their pronunciation if they could find ready at hand some small but adequate book of reference.

In response to the demand of these multitudes, I have prepared, with some misgivings and hesitation, this handy pronouncing guide. It includes not only a host of the most commonly used and often mispronounced technical terms, but also the better known generic names of plants and animals and numerous Latin adjectives and adjectival compounds used as specific or trivial names. As a special aid to learning, with each specific name is given its original Greek or Latin meaning or English equivalent.

Acceptable pronunciation of each word is indicated, by its division into parts (not necessarily syllables) by means of hyphens, by accent, and by diacritical marks. The preferred pronunciation is, in each case, indicated first; the less used but acceptable alternatives are placed afterwards.

It should ever be remembered that while there are formal rules of pronunciation they have not always been observed. Long usage has in certain cases established other ways of sounding some letters, especially vowels, and of placing accents. It is also well to keep in mind that words, especially derived ones, may be pronounced differently by phonetic experts and by reputable biologists residing in different countries. The individual preferences are indeed many.

The scientific names of both plants and animals are generally cast in Latin form even though they may be compounded from Greek or other stems. It has been agreed that they should, for the most part, be pronounced in conformity to Latin rules and practice. Accordingly I have given in the introduction the more important rules governing the syllabification and accentuation of Latin words. A table showing the needed diacritical marks and the sounds of the letters which they govern has been located for ready reference inside both the front and back covers of the book. To add interest and to help the student in learning, a number of illustrations have been placed throughout the text. These call attention to often mispronounced words.

Those who use this Guide are urged to read carefully the Introduction and to make an earnest effort to master the brief but highly important material found there. They may then proceed to pronounce intelligently and with ease and accuracy, the names and terms they use.

Gardeners and horticulturists, specialists in animal husbandry, foresters, naturalists, and students of the biological sciences will often need to consult the Guide. Teachers, especially, will welcome this aid as they prepare to give their lectures or conduct recitations. All will find it to be a most profitable and interesting spare-time pleasure (although at times embarrassing) to run through the lists of familiar generic and specific names as well as oft-used technical terms and underline the great number of mispronunciations they have been habitually making.

In preparing the long list of words (there are more than 9000), a guide to whose pronunciation is indicated, the author tried particularly to include only those most likely to cause difficulty. Thus many commonly used terms and generic names of obvious sound and accentuation have been omitted. This has made it possible to keep the book down to a size easy to handle, and to render it valuable as a constant desk or brief-case companion.

The author realizes that although he has spared no effort to here present a thoroughly reliable work, there still must be errors which have slipped in. Constructive criticisms and corrections are accordingly invited for the purpose of helping to attain to a more uniform and correct standard of pronunciation in future editions of the HANDBOOK.

I have been fortunate in securing the advice and generous assistance of many able students of the principles of phonetics, and to them I am deeply indebted. A list of the more important and helpful books consulted is appended.

EDMUND C. JAEGER

Riverside, California

INTRODUCTION

UNFORTUNATE AS IT IS, the "English method" of pronouncing Latin is used, both in this country and in England, by most biologists, in the pronunciation of biological names cast in Latin form. According to this system the vowels are given their customary English sounds as are the consonants, except that ch is pronounced as k; c, g, and gg are usually soft before e, i, y, and the diphthongs ae and oe. The usual Latin rules of accentuation are observed.

The use of the "English method" of pronunciation goes back to the period when this method was used in the English Law Courts. Later it was widely taught in English and American schools. More recently it has been superseded by the "Continental method," which is now used exclusively in the secondary schools and colleges of the United States and many parts of Europe.

Since there are those who may prefer to pronounce words in accordance with the Continental or Roman method the following explanation of diacritical marks and sounds of consonants and diphthongs will be useful:

Long Vowels

ā like a in ah
ē " e " they
ī " i " machine
ō " o " mole
ū " u " mule

Short Vowels

a like a in idea
e " e " let
i " i " pin
o " o " obey
u " u " full

Consonants

c like c in come
ch " ch " chemistry
g " g " give
r " r " room
s " s " son
t " t " time
v " w " we
qu " qu " quite

Diphthongs

ae nearly like ai in aisle
oe " " oi " coin
au " " ou " spout
eu " " eu " feud
ei " " ei " veil
vi " " we " cui (kwe)

CONCERNING THE SYLLABIFICATION AND ACCENTUATION OF LATIN WORDS OR OF GREEK WORDS CAST IN LATIN FORM

1. A syllable consists of a vowel or diphthong* with or without one or more consonants. Accordingly, a word has as

    many syllables as it has separate vowels or diphthongs.
2. In dividing a word into syllables, a single consonant is joined to the vowel which follows it.
3. If two or more consonants occur between two vowels, as many are joined to the following vowel as can be pronounced with it.
4. In compounds, the parts are separated.
5. The last syllable of a word is called the *ulltima*. The next to the last syllable of a word is called the *penult*. The syllable preceding the penult is called the *antepenult*.

Words of two syllables have the accent on the penult. Thus: Latin *tū'-bą*, trumpet and *ăn'-cĕps*, two headed, double.

Words of more than two syllables have the accent on the penult *when that syllable is long*: otherwise the accent falls on the antepenult. Thus: *prae-dī'-cō*, to foretell but *prae'-dĭ-cō*, to declare.

In this pronouncing guide only the primary or principal accent is indicated, since, usually, knowing this, it is rather easy to find the secondary accent. It is well to remember that the secondary accent, as a general rule, can never fall less than two syllables before the primary one.

A syllable is long:

1. if its vowel is long. In this book the long vowels are marked; unmarked vowels must be regarded as short.

---

    * A diphthong (Gr. *di*, double; *phthongos*, voice) is a union of two vowels pronounced as one. In "proper" diphthongs, which we have in such English words as "joy," "poise," and "round," the two vowels are blended; but in "improper" diphthongs which appear in such words as "people," "each," and "pain" only one of the vowels, generally the first, is sounded.

2. if its vowel is followed by "x" or "z."
3. if its vowel is short but followed by two or more consonants. Except a mute (p, b, t, d, c, k, g, q,) followed by l or r or by x or z.
4. if it contains a diphthong.*
5. final *as*, *es*, *os* are long.

A syllable not held to be long is short.

A final syllable ending in any consonant other than "s" is short (-is, -us, and -ys are short). There are few exceptions.

In pure Latin words a vowel is long:

1. if it is formed by the contraction of a diphthong.
2. if it occurs before "gm" (and often "gn"), "nf," and "ns."
3. if it occurs before "consonant i" = y (with the sound of *y* in *yet*).
4. usually if the "o" and "u" are final.

In pure Latin words a vowel is short:

1. if it occurs before another vowel or "h," thus: *via, nihil.* An exception is found in some words transcribed from Greek.
2. if it occurs before "nd" or "nt." Thus: *amandus* and *amant.*

Compound Words

Many generic and trivial (specific) names of animals and plants consist of fabricated compound

---

* The most common Latin diphthongs are "ae," "au," and "oe." "eu" occurs in a few Latin words derived in part from the Greek "eu," meaning "well, good." "-eus," a common Latin adjectival ending, is pronounced "-ĕ-us"; i.e., in two syllables. The common Greek diphthongs "αε" (represented by ae), "ευ" (transliterated "eu"), and "οι" (transliterated "oe") are considered long in pronunciation.

words. If the words are compounded from Greek words or word-stems, the parts are often joined by the use of the vowel "o," often, but not necessarily, sounded as a shortened long "o" and marked "ŏ." Thus we have *ornithopteris* from the Greek stem *ornithos*, a bird, joined by the use of "o" to the word *pteris*, a wing. This connecting vowel "o" is also used in constructing some Latin compounds but the usual joining vowel in Latin compounds is "i." This we see in the trivial name *alnifolia* from the Latin stem *alnus*, the alder, and *folium*, a leaf.

Transliterated Word-Endings

Greek words ending in -*on* (-ον) and -*os* (ος), when made over into Latin words, appear with the endings -*um* and -*us*, while those ending in long *e* (-η) usually have their Latin derivatives ending in -*a*. Thus Greek *petalon* (πεταλον) becomes the Latin *petalum*, a leaf, and Greek *cyamos* (κύαμος) becomes the Latin generic name *Cyamus*, a bean.

In the case of commemorative names ending in -*ia*, -*iana*, or occasionally in -*ella*, given to honor discoverers, eminent scholars, or patrons of science, every effort should be made to preserve in their pronunciation as near as possible the original sounds; only thus can the names be readily associated with the persons in whose memory they were originally given. Certainly Dahlia, given to commemorate the Swedish botanist Dahl, should be pronounced Dä′-lĭ-ạ and not Dă′-lĭ-ạ as is so commonly done, and Camellia, given in honor of the botanical contributions of George Joseph Kamel (Latinized form = Camelli), the seventeenth century Moravian traveller, should have the *e* pronounced short (Căm-ĕl′-lĭ-ạ) and not long (Căm-ēl-′lĭ-ạ) as so many careless persons are in the habit of doing. Pronounced otherwise, the connection between the man and the plant or animal is almost entirely obliterated and one of the chief purposes of giving the name is defeated.

Sometimes words have passed over into the English language and in so doing have not only had their accent shifted and the sounds of their vowels changed, but they have also had the spelling slightly altered. Examples of such words are the Greek ănĕm-ō′-nē (ἀνεμώνη) which in English appears as the plant name anĕm-ō-nē, and the Latin or-ā′-tor becomes in English ôr′-ā-tôr. From the Latin fŏ-li-us, leaf, we have the English words fŏ′-lĭ-āge and fŏ′-li-ō. Strange indeed, and rightly so, it now would sound, to hear someone speak of fŏ′-lĭ-āge or of a fŏ′-li-ō. The long o in foliage comes to us through French.

Generic names of plants are followed by an asterisk (*).

A

Abama* (ab-ā'-mạ)
abbreviatus (ab-rev-i-ā'-tus) abridged.
abditus (ab'-di-tus) removed, withdrawn.
abdomen (ab-dō'-men)
abductor (ab-duk'-tôr)
Abelia* (āb-el'-i-ạ; ā-bēl'-i-ạ)
aberrant (ab-er'-ant)
abient (ab'-i-ent)
Abies* (ab'-i-ēz)
abieticola (ab-i-et-i'-kô-lạ) fir-dweller.
Abietineae* (ab-i-et-in'-ê-ē)
abietinus (ab-i-et-ī'-nus) abies-like.
Ablepharus (á-blef'-á-rus)
ablutus (ab-lū'-tus) washed.
abnuitus (ab-nu'-it-us) given up, rejected.
abomasum (ab-ô-mā'-sum)
aboral (ab-ō'-ral)
aboriginus (ab-ôr-ij'-in-us) the primeval Romans, also, a nation, the Aborigines.
abortivus (a-bôr-tī'-vus) born prematurely.
Abramis (ab'-rá-mis)
abrasus (ab-rā'-sus) rubbed off, shaved.
Abraxas (á-brak'-sas)
Abrocoma (á-brok'-ô-mạ)
Abronia* (á-brō'-ni-ạ)
abrotanifolius (ab-rot-an-i-fol'-i-us, ab-rot-an-i-fō'-li-us) with leaf like *Abrotanum*.

abrotanoides (ab-rot-an-o-īd'-ēz) like *abrotanum* or southernwood.

Abrotanum* (ab-rot'-an-um)

abrotonoides (ab-rot-on-o-ī'-dēz) resembling *Artemisia* (*abrotonum*).

abruptus (ab-rupt'-us) separated, torn off.

Abrus* (ā'-brus)

abscise (ab-sīz')

abscissus (ab-sis'-us) separated, divided.

absconsus (ab-skon'-sus) concealed, hidden away.

absimilis (ab-sim'-il-is) unlike.

absorb (ab-sôrb')

absorption (ab-sôrp'-shun)

abutiloides (ab-ū-til-o-īd'-ēz) like *Abutilon*.

Abutilon* (ab-ū'-ti-lon)

Acacia* (ȧk-ā'-shi-ạ)

Acaena (a-sē'-nạ)

Acaloithus (ak-al-ȯ-ith'-us)

Acalypha* (a-kȧ-lī'-fạ)

Acanthaceae* (ak-an-thā'-sê-ē)

Acanthina (ȧk-anth'-in-ạ)

Acanthis (ȧk-anth'-is)

Acanthium* (ȧk-anth'-i-um)

acanthocoma (ȧk-anth-ok'-om-ạ) spiny-haired.

acanthodes (ȧk-anth-ō'-dēz) thorny.

Acanthodoris (ȧk-anth-ȯ-dō'-ris)

Acanthodrilus (ȧk-anth-ȯ-drī'-lus)

acanthoides (ȧk-anth-o-ī'-dēz) like a thorn or like *Acanthus*.

Acantholimon* (ȧk-anth-ȯ-lī'-mon)

Acanthophis (ȧk-anth'-ȯ-fis)

Acanthostachyum* (ȧk-anth-ost-ak'-i-um)

Acarida (a-kâr'-id-ạ)
Acarospora* (ak-âr-os'-pô-rạ)
Acarus (ak'-âr-us)
acaulescent (ạ-kôl-es'-ent)
acaulis (ạ-kô'-lis) without stem.
accedens (ak-sē'-denz) resembling, approaching.
Accipiter (ak-sip'-i-têr)
accipitrine (ak-sip'-it-rin, ak-sip'-i-trīn)
acclimatize (ạ-klī'-ma-tīz)
acclinis (ak-lī'-nis) leaning on or against some-
    thing.
acephalus (ạ-sef'-al-us) without head.
Acer* (as'-êr)
acer (ā'-sêr) with sharp taste, sharp.
Aceraceae* (as-e-rā'-sê-ē)
Aceras* (ā'-ser-as)
acerate (as'-ê-rāt)
Acerates* (as-êr-ā'-tēz)
acerbus (as-êrb'-us) bitter, harsh, rough.
Acerentomidae (ās-er-en-tom'-i-dē)
acerifolius (as-er-i-fol'-i-us, as-er-i-fō'-li-us) maple-
    leaved.
aceris (as'-er-is) of the maple tree.
acerosus (as-er-ō'-sus) needle-shaped, full of
    needles; also, chaffy.
acervate (ạ-sêr'-vāt, ạs'-er-vāt)
acetic (ạ-sē'-tik; ạ-set'-ik)
acetosus (as-ê-tō'-sus) full of acid.
Achatina (a-kat'-in-ạ)
achatinus (ak-ā-tī'-nus, ak-at-ī'-nus) like agate, of
    the color of agate.
achene (ạ-kēn'; ak-ēn')

Achillea* (ak-i-lē′-ạ)
achilleaefolius (ak-il-ē-ē-fol′-i-us, ak-il-ē-ē-fō′-li-us)
  with leaf like yarrow.
Achimenes* (ạ-kim′-e-nēz)
Achlys* (ak′-lis)
achradophilus (ak-rad-of′-i-lus) pear-tree loving.
Achras* (ak′-ras)
Achyronia* (ak-ir-on′-i-ạ)
Achyropappus* (ak-ir-ŏ-pap′-us)
acicularis (ạ-sik-ul-ā′-ris) needle-like.
Acidanthera* (as-id-an′-thĕ-rạ)
acidophil (as′-id-ŏ-fil, as-id′-ŏ-fil)
Acilius (as-il′-i-us)
acinacifolius (ạ-sin-ā-si-fol′-i-us, ạ-sin-ā-si-fō′-li-us)
  with sword-shaped leaf.
acinaciform (ạ-sin-ạ′-si-fôrm)
acinarius (as-in-ā′-ri-us)
Acineta (as-in-ē′-tạ)
acinifolius (as-in-i-fol′-i-us, as-in-i-fō′-li-us) having
  leaves resembling those of *Acinos arvensis*.
Acinonyx (ạs-in′-on-iks)
acinus (as′-in-us, pl. as′-in-ī)
acipenser (a-sip-en′-sêr)
Aciphylla* (as-iph-il′-ạ)
Acisanthera* (as-is-an-thē′-rạ)
Acmadenia* (ak-mad-ē′-ni-ạ)
Acmena* (ak-mē′-nạ)
Acnida* (ak-nī′-dạ)
Acocanthera* (ak-ō-kan-thē′-rạ)
Aconitum* (ak-ŏ-nī′-tum)
acontium (ạ-kon′-shi-um, ạ-kon′-ti-um)
Acordulecera (ak-ôrd-ul-ēs′-er-ạ)

Acorus* (ak'-ô-rus)
acoustic (à-kōōs'-tik)
Acradenia* (ak-ra-dē'-ni-ą)
Acraspeda (à-kras'-pe-dą)
Acridotheres (a-krid-ô-thē'-rēz)
Acris (āk'ris)
Acrocera (ak-ros'-êr-ą)
Acroceratidae (ak-rô-se-rat'-i-dē)
Acrochordinae (ak-rô-kôr-dī'-nē)
Acrochordus (ak-rô-kôrd'-us)
Acrocinus (ak-rô-sī'-nus)
acrocladon (ak-rok'-lad-on) with pointed branch
Acrocomia* (ak-rô-kō'-mi-ą)
acrogynous (ak-roj'-i-nus)
Acrolepia (ak-rol-ē'-pi-ą)
Acronychia* (ak-ron-ik'-i-ą)
Acronycta* (ak-ron-ik'-tą)
Acropera* (ak-rop-ē'-rą)
Acrosanthes* (ak-ros-anth'-ēz)
Acrosoma (ak-ro-sō'-mą)
acrostical (ak-ros'-ti-kal)
Acrostichum* (ak-ros'-tik-um)
acroteric (ak-rô-tē'-rik)
acrotrichus (ak-rot'-rik-us) with sharp hairs.
Acryllium (ak-ril'-i-um)
Actaea* (ak-tē'-ą)
Actinella (ak-ti-nel'-ą)
Actinemys (ak-tin'-em-is)
Actinocarpus* (ak-ti-nok-âr'-pus)
Actinolepis* (ak-ti-nol'-ep-is)
Actinomeris* (ak-ti-nom'-êr-is)
Actinophrys (ak-ti-nof'-ris)

Actinostachys (ak-ti-nost'-ak-is)
Actinostrobus* (ak-ti-nost'-rob-us)
actius (ak'-ti-us) pertaining to *Actium*.
Acuan* (ak'-û-an)
aculeatus (ak-ū-le-ā'-tus) thorny, prickly.
acuminatus (ak-ū-min-ā'-tus) sharpened, pointed.
acus (ak'-us) a pin or needle, something pointed.
acutangulus (ak-ū-tang'-ul-us) with sharp or well-defined angles.
acutiflorus (ak-ū-ti-flō'-rus) with sharp-pointed flowers.
acutifolius (ak-ū-ti-fol'-i-us, ak-ū-ti-fō'-lí-us) with sharp-pointed leaves.
acutipennis (a-kū-ti-pen'-is) sharp-feathered, pointed-feathered.
acutus (ak-ū'-tus) acute or pointed.
Adela (ad-ē'-lạ)
Adelea (ad-ē-lē'-ạ)
Adelges (à-del'-jēz)
Adelgidae (à-del'-ji-dē)
Adelochorda (ad-ēl-ô-kôr'-dạ)
adelphogamy (à-del-fog'-a-mi)
adeniform (à-dē'-ni-fôrm; à-den'-i-fôrm)
adenocaulon (à-dē-no-kôl'-on) glandular-stemmed.
Adenophora* (a-dē-nof'-ôr-ạ)
adenophyllus (à-dē-nof-il'-us) glandular-leaved.
adenose (a'-dê-nōs)
Adenostoma* (a-dē-nost'-o-mạ)
Adenota (a-dē-nō'-tạ)
Adenotrichia* (a-dē-nô-trik'-i-ạ)
Adephaga (à-def'-à-gạ)
adiantifolius (ad-i-ant-i-fol'-i-us, ad-i-ant-i-fō'-li-us) with leaf like *Adiantum*.

Adiantum* (ad-i-an'-tum)
adience (ad'-i-ens)
Adimeridae (ad-im-er'-id-ē)
adipocere (ad'-ip-ô-sēr)
adipose (ad'-i-pōs)
adjectus (ad-jek'-tus) placed near.
Adlumia* (ad-lū'-mi-ạ)
adminiculatus (ad-min-ik-kul-ā'-tus) well-supported, well furnished.
admirabilis (ad-mīr-ā'-bil-is) worthy of admiration, wonderful.
adnascens (ad-nas'-senz) growing on.
adnatus (ad-nāt'-us) growing to, connected by birth.
adnixus (ad-niks'-us) leaning upon, pressing against.
adocetus (ad-os-ē'-tus) unexpected.
Adonidia* (ad-on-id'-i-ạ)
Adonis* (ad-ō'-nis)
Adoxa* (ȧ-dok'-sạ)
adpressus (ad-pres'-us) pressed toward.
adrenal (ad-rē'-nal)
adscendens (ads-sen'-denz) growing up, standing higher.
adsitus (ad'-si-tus) sown, planted, set near something.
adspersus (ad-spêrs'-us) a sprinkling upon, scattering.
adsurgens (ad-sūr'-jenz) rising, erect, standing up.
adtidal (ad-tī'-dal; ad'-tĭd-al)
adulterinus (ad-ul-ter-ī'-nus) not genuine.
adumbratus (ad-umb-rā'-tus) false.
aduncus (ad-unk'-us) hooked, bent like a hook.

adustus (ad-ust'-us) burned, singed, damaged, made brown.

advenus (ad'-ven-us) a stranger, a foreigner.

Aechmea* (ēk-mē'-ạ; ēk'-me-ạ)

Aechmophorus (ēk-mof'-ȯ-rus)

aecidiospore (ē-sid'-i-ȯ-spôr)

aecidium (ē-sid'-i-um)

aeciospore (ē'-si-ȯ-spôr)

aedeagus (ē-dē'-ag-us; ēd-ē-ā'-gus)

aedon (a-ē'-don) the nightingale.

aedonius (a-ē-don'-i-us) pertaining to the nightingale.

Aega (ē'-gạ)

aegaeus (ê-jē'-us) Aegean.

aegagrus (ē-gā'-grus) the wild goat.

aeger (ē'-ger) troubled, suffering, sick.

Aegeriidae (ē-je-rī'-i-dē)

Aegialites (ē-ji-al-ī'-tēz)

Aegicerus* (ē-jis'-er-us)

Aegilops* (ē'-jil-ops)

Aegithalos (ē-ji-thā'-los)

Aegopodium* (ē-gȯ-pō'-di-um)

Aegopogon* (ē-gȯ-pōg'-ōn)

Aegothelidae (ē-gȯ-thel'-i-dē)

aegrotus (ē-grō'-tus) sick; also, sick of seeing you.

Aeluropus (ē-lū'-rȯ-pus)

aemulus (ē'-mul-us) rivalling.

aeneus (ē'-ne-us) of bronze or copper.

Aeolidiadae (ē-ȯ-li-dī'-ad-ē)

Aeolothripidae (ē-ȯ-lō-thrip'-i-dē)

Aeonium* (ē-ō'-ni-um)

Aepyceros (ē-pis'-êr-os)

Aepyornis (ēp-i-ôr′-nis)
aequabilis (ē-kwā′-bi-lis) equal, similar.
aequipetalus (ē-qui-pet′-al-us) equal-petalled.
Aequorea (ē-kwō′-rē-ạ)
aequoreus (ē-kwôr′-e-us) of or pertaining to the sea.
aereus (ē′-re-us) coppery, made of copper or bronze.
aërial (ā-ē′-ri-al, a-ē′-ri-al)
Aërides* (ā-ē′-ri-dēz)
aërius (ā-ē′-ri-us) pertaining to air, high, transitory.
Aërobion* (ā-ēr-ōb′-i-on)
aërotropic (ā-êr-ô-trop′-ik)
aeruginosus (ē-rū-jin-ōs′-us) full of copper, rusty.
Aeschna (ēsk′-nạ)
Aeschnidae (esk′-ni-dē)
Aeschynanthus* (es-ki-nan′-thus)
Aeschynomene* (es-ki-nom′-ê-nē)
aesculifolius (es-kul-i-fol′-i-us, es-kul-i-fō′-li-us) oak-leaved.
Aesculus* (es′-ku-lus)
aestival (ēs′-ti-val; ēs-tī′-val)
aestivalis (ēs-ti-vā′-lis) summer-flowering, pertaining to summer.
aestivate (ēs′-ti-vāt)
aestivation (ēs-ti-vā′-shun)
aestivus (ē′-sti-vus) of summer; often, in botany, referring to time of flowering.
aestuans (ēs′-tu-anz) warming, inflaming.
aethereus (ē-thē′-ri-us) heavenly.

Aethionema* (ē-thi-on-ē'-mạ)
aethiopicus (ē-thi-ō'-pik-us) from Ethiopica (Aethi-
  opica).
Aethusa* (ē-thū'-sạ)
aetiology (ē-ti-ol'-ôj-i)
aetites (ā-e-tī'-tēz)
afer (āf'-êr) African.
afferent (a'-fer-ent)
affinal (af-ī'-nal)
affinis (af-īn'-is) adjacent, neighboring.
afrum (ā'-frum) African.
Agabus (ag'-ạ-bus)
Agalmyla* (ag-al'-mil-ạ)
Agama (a'-gam-ạ)
Agamidae (ȧ-gam'-i-dē)
agamospecies (ag-am-ȯ-spē'-shēz)
agamospore (ag-am'-ȯ-spôr)
Agaontidae (ag-ȧ-on'-ti-dē)
Agapanthus* (ag-ap-anth'-us)
agape (ȧ-gāp', ȧ-gap')
Agapostemon (ag-ap-os'-te-mon)
Agaricus (ag-ar'-i-kus)
Agastachys* (ȧg-ast'-ak-is)
Agathaea* (ag-ath-ē'-ạ)
Agathis* (ag'-ȧ-this)
Agathophyllum* (ag-ath-of-il'-um)
agave (ȧ-gä'-vê)
Agave* (ȧ-gā'-vē; ȧ-gä'-vē)
agavoides (ag-āv-o-ī'-dēz; ag-äv-o-ī'-dēz) like
  Agave.
Agdestis* (ag-des'-tis)
Agelaius (aj-ê-lā'-yus)

Agelena (aj-ē-lē′-nạ)
Ageniapsis (aj-en-i-as′-pis)
ageratoides (aj-ē-rat-o-ī′-dēz) like ageratum.
Ageratum* (aj-ē′-rat-um, a-jêr′-a-tum)
aggregatus (ag-rē-gā′-tus) gathered together.
agilis (a′-ji-lis) active, nimble.
Agkistrodon (ag-kis′-trô-don)
Aglaonema* (ag-lȧ-on-ē′-mạ)
Agnepteryx (ag-nep′-ter-iks)
agninus (ag-nī′-nus) pertaining to a lamb.
Agnostus* (ag-nō′-stus)
Agoseris* (ag-os′-êr-is)
Agraphis* (ag′-raf-is)
agrarius (ag-rā′-ri-us) pertaining to fields.
agrestis (a-gres′-tis) wild.
agrifolius (ag-ri-fol′-i-us, ag-ri-fō′-li-us) rough or
    scabby-leaved.
Agriidae (ag-rī′-i-dē)
Agrilus* (ag′-ril-us)
Agrimonia* (ag-ri-mō′-ni-ạ)
Agriotes* (ag-ri-ō′-tēz)
Agromyzidae (ag-rô-mīz′-i-dē)
Agropyron* (ag-rô-pī′-ron)
Agrostemma* (ag-rô-stem′-ạ)
agrostideus (ag-rōs-ti′-de-us) like agrostis, a plant
    mentioned by Theophrastus.
Agrostis* (ag-rōs′-tis)
Agrotis (a-grō′-tis)
Agulla (ag-u′-lạ)
aigrette (ā-gret′, ā′-gret)
ailanthifolius (ā-lanth-i-fol′-i-us, ā-lanth-i-fō′-li-us)
    with leaves like *Alianthus*.

Ailanthus* (ā-lan'-thus; ī-lan'-thus)
Ailurin (ī-lūr'-in)
Ailuroedus (āl-ū-rē'-dus)
Ailuropoda (āl-ū-rop'-ȯ-dạ)
Ailuropus (āl-ū-rō'-pus)
Aimophila (ī-mof'-il-ạ)
Aiphanes* (ā-ī'-phan-ēz)
Aira* (ā'-rạ)
aithochroi (ī-thok'-rȯ-ī)
aitionastic (ī-ti-ȯn-as'-tik)
Aix (āks)
aizoides (ā-ī-zo-ī'-dēz) aizoon-like.
aizoon (ā-ī-zō'-on) ever-living; an evergreen plant.
Ajaia (ī-ī'-a)
ajaja (ī-ī'-ạ; ä-yä'-yä)
Ajuga* (aj'-o͞o-gạ; aj-ū'-gạ)
akebia (ak-ē'-bi-ạ)
akinesis (ak-in-ē'-sis)
akinete (ak'-i-nēt)
alacer (al'-a-ser) quickly, lively.
alatavicus (al-at-av'-ik-us)
alate (ā'-lāt)
alatus (ā-lā'-tus) winged.
Alauda (a-lȯ'-dạ)
Alaus (al-ā'-us)
albatross (al'-bạ-tros)
albescens (al-bes'-senz) growing white.
albicans (al'-bi-kanz) becoming white.
albicaulis (al-bik-ȯ'-lis) white-stemmed.
albicollis (al-bik-ol'-is) white-necked.
albidulus (al-bid'-ul-us) whitish.
albidus (al'-bi-dus) white.

albifrons (al'-bif-ronz) white-browed.
albigula (al-bi-gū'-lạ)
albigulus (al-bi-gū'-lus) white-throated.
albinism (al'-bi-nizm)
albino (al-bī'-nō)
albipes (al'-bi-pēs) white-foot.
albispinus (al-bis-pī'-nus) white-spined.
Albizia* (al-biz'-i-ạ)
albocinctus (al-bō-sink'-tus) white-belted.
albostipes (al-bō-stī'-pēz) white-stalked, with white stalk.
albulus (al'-bul-us) whitish.
albumen (al-bū'-men)
albus (al'-bus) white.
Alca (al'-kạ)
Alcea* (al'-se-ạ)
Alcedo (al-sē'-dō)
Alcelaphinae (al-sel-ạ-fī'-nē)
Alcelaphus (al-sel'-ạ-fus)
Alces (al'-sēz)
Alchemilla* (al-kê-mil'-ạ)
alcicorneus (al-si-kôr'-ne-us) antler-shaped, with horns like the elk.
alcoides (al-ko-ī'-dēz) auk-like.
Alcyonium (al-si-ō'-ni-um)
Alectoria* (a-lek-tō'-ri-ạ)
Alectoris (a-lek'-tôr-is)
Alectrion (a-lek'-tri-on)
Alectura (a-lek-tū'-rạ)
Aleochara (al-ê-ok'-ạ-rạ)
aleppensis (al-ep-en'-sis) from Aleppo.
Aletris* (al-et'-ris, al'-ê-tris)

aletroides (al-et-ro-ī'-dēz) like *Aletris*.
Aleurobius (al-û-rō'-bi-us)
aleuron (al-ū'-ron)
Aleurodidae (al-û-rod'-i-dē)
alga (al'-gạ, pl. al'-jē)
algidus (al'-ji-dus) cold.
alimentary (al-i-men'-ta-ri)
Alisma* (ạ-liz'-mạ)
alismaefolius (al-iz-mē-fol'-i-us, al-is-mē-fō'-li-us)
    with leaves like *Alisma*.
alkaline (al'-kạ-līn; al'-kạ-lin)
alkeifolius (al-ke-i-fol'-i-us, al-ke-i-fō'-li-us) with
    leaves like mallow.
Allactaga (ạ-lak'-tạ-gạ)
Allamanda* (al-ạ-man'-dạ)
allantoic (al-an-tō'-ik)
allantois (a-lan'-tô-is)
Alle (al'-ē)
allele (ạl-ēl', pl. ạ-lēlz')
allelism (ạ-lē'-lizm)
allelomorph (ạ-lē'-lô-môrf)
Allenrolfea* (al-en-rol'-fe-ạ)
allex (al'-eks) the great toe.
alliaceus (al-i-ā'-se-us) garlic-like.
Alliaria* (al-i-ā'-ri-ạ)
Allium* (al'-i-um)
allochthonous (al-ok'-thôn-us)
allogamus (al-og'-ạ-mus)
allometry (al-om-et'-ri)
allophyllus (al-of-il'-us) with other kind of leaves,
    i.e., with peculiar, strange leaves.
Allosaurus (al-ô-sôr'-us)

allosome (al'-ô-sōm)
Allosorus* (al-os-ō'-rus)
allotropic (al-ô-trop'-ik)
almond (ä'-mund)
alnifolius (al-ni-fol'-i-us, al-ni-fō'-li-us) with leaves like the alder.
Alnus* (al'-nus)
Alocasia* (al-ô-kā'-shi-ą)
aloe (al'-ō, pl. al'-ōz)
Aloe* (al'-o-ē)
alogus (a'-lô-gus) irrational, without reason.
aloides (al-o-ī'-dēz) resembling *Aloe.*
aloifolius (al-o-i-fol'-i-us, al-o-i-fō'-li-us) aloe-leaved.
Alonsoa* (al-on-sō'-ą)
Alopecias (al-ô-pē'-shi-as)
alopecuroides (al-ô-pek-ū-ro-ī'-déz) foxtail-like.
Alopecurus* (al-ô-pek-ū'-rus)
Alopex (al'-ô-pēks)
Alosa (ȧ-lō'-są)
Alouatta (al-ŏŏ-at'-ą)
Aloysia* (al-ô-ish'-i-ą)
alpestris (al-pest'-ris) of the Alps.
alpine (al'-pīn, al'-pin)
alpinus (al-pī'-nus) alpine.
Alsine* (al-sī'-nē)
alsinifolius (al-sin-i-fol'-i-us, al-sin-i-fō'-li-us) like *Alsine,* the chickweed.
Alsophila (al-sof'-il-ą)
Alstroemeria* (al-strē-me'-ri-ą)
alternans (al-têrn'-anz) changed.
alternate (al-têr'-nat, al'-ter-nāt)

alterniflorus (al-têrn-i-flōr'-us) with alternating flowers.
Althaea* (al-thē'-ạ)
alticolus (al-tik'-ŏl-us) dwelling in high places.
Altides (al-tī'-dēz)
altilis (alt'-il-is) nutritive, fat, large.
altipetens (al-tip'-et-enz) seeking high places.
altissimus (al-tis'-im-us) tallest, very tall.
altiusculus (al-ti-us'-ku-lus) rather high, a little too high.
altivallis (al-ti-val'-is) of high valleys.
altrices (al'-tri-sēz)
altricial (al-tri'-shal)
altus (al'-tus) high, tall.
alula (al'-ū-lạ)
alumnus (al-um'-nus) well-nourished, flourishing.
alutacius (a-lū-tā'-shi-us) pertaining to soft leather.
alvarius (al-vā'-ri-us) pertaining to or of the womb.
alveolar (al-vē'-ŏ-lêr; al'-vĕ-ŏ-lâr)
alveolus (al-vē'-ŏ-lus, al-ve'-ol-us)
alveus (al'-ve-us) a basket, a deep hollow, a channel.
Alydus (al'-i-dus)
Alysicarpus* (al-is-i-kâr'-pus)
Alyssum* (a-lis'-um)
Alytes (al'-i-tēz)
amabilis (ȧ-mā'-bi-lis) lovely, worthy of love.
Amanita* (am-ȧn-ī'-tạ)
amanous (am'-ȧ-nus)
Amaranthus* (am-ȧ-ran'-thus)
Amaroucium (am-âr-ū'-shi-um)
amarus (a-mā'-rus) bitter.

Amastridium (a-mas-trid′-i-um)
ambiguus (am-big′-u-us) doubtful, of uncertain relationship.
ambitus (am′-bi-tus) a going round, a revolving.
Ambloplites (am-blop-lī′-tēz)
amblyceps (am′-bli-seps) blunt-head.
Amblychila (am-bli-kī′-lạ)
amblyodon (am-blī′-od-on) blunt-toothed.
ambon (am′-bon)
ambrosia (am-brō′-zhi-ạ, am-brȯ′-zi-ạ)
ambulacral (am-bū-lāk′-ral)
ambulacrum (am-bū-lāk′-rum)
Ambystoma (am-bis′-tȯ-mạ)
ameiosis (ạ-mī-ō′-sis)
Amelanchier* (am-ê-lan′-ki-êr)
ameloblast (a-mel′-ȯ-blast)
ament (ā′-ment, am′-ent)
americanus (am-er-ik-ā′-nus) of America.
amethysteus (am-eth-is′-te-us) like amethyst.
Ametropodidae (am-e-trȯ-pod′-i-dē)
Amianthium* (am-i-anth′-i-um)
amine (am′-in, am′-ēn)
amino (a-mē′-nō; am′-i-nȯ)
Ammocharis* (am-ok′-ar-is)
ammocoete (am′-ȯ-sēt)
ammocoetes (am-ȯ-sē′-tēz)
Ammodramus (am-od′-ra-mus)
Ammodytes (am-ȯ-dī′-tēz)
Ammogeton* (am-og-ē′-ton)
Ammon (am′-on) an epithet of Zeus.
Ammonites (am-ȯ-nī′-tēz)
Ammophila* (am-of′-ilạ)

ammophilus (am-of'-il-us) sand-loving.
Ammospermophilus (am-ô-spûr-mof'-il-us)
amnion (am'-ni-on)
amoebiasis (a-mê-bī'-à-sis)
Amoebina (a-mē-bī'-nạ)
amoeboid (a-mē'-boyd)
amoenus (a-mē'-nus) lovely, charming.
Amomum* (am-ō'-mum)
Amoreuxia (am-ô-rŏŏk'-shi-ạ)
Ampelopsis* (am-pe-lop'-sis)
Amphiachyris* (am-fi-a'-kir-is)
amphibious (am-fib'-i-us)
amphiblastula (am-fi-blas'-tŭ-lạ)
Amphibolurus (am-fi-bol-ūr'-us)
Amphicarpaea* (am-fi-kâr-pē'-ạ)
Amphicarpum* (am-fi-kârp'-um)
Amphicepha (am-fi-sē'-fạ)
Amphicerus (am-fi'-ser-us)
amphicoelus (am-fi-sē'-lus)
Amphicyon (am-fis'-i-on)
Amphilobium* (am-fil-ob'-i-um)
Amphimeryx (am-fi'-mer-iks)
amphimixis (am-fim-ik'-sis)
Amphipoda (am-fip'-ô-dạ)
amphipodous (am-fip'-o-dus)
Amphisbaena (am-fis-bē'-nạ)
Amphispiza (am-fis-pī'-zạ)
amphithecium (am-fi-thē'-shi-um)
Amphithoë (am-fith'-ô-ē)
amphitriaene (am-fit-rī'-ēn)
amphitropous (am-fit'-rop-us)
Amphiuma (am-fi-ū'-mạ)

Amphiura (am-fi-ūr'-ạ)

Amphizoidae (am-fi-zō'-i-dē)

amphrysus (am-frī'-sus) of Amphrysos, a river in Thessaly.

amplexicaulis (am-pleks-i-kôl'-is) with entwining or embracing stem.

ampliate (am'-pli-āt)

amplus (am'-plus) great, large, wide.

ampulla (am-pōōl'-ạ), am-pul'-ạ)

ampullaceus (am-pul-ā'-se-us) flask-like.

ampullatus (am-pul-ā'-tus) jugged, bottled.

Amycterus (ạ-mik'-tĕ-rus)

Amyda (am'-id-ạ)

amygdalinus (am-ig-dal-ī'-nus) of almonds.

amygdaloides (am-ig-dal-o-ī'-dēz) like an almond.

Amyris* (am'-i-ris)

Anabas (an'-ạ-bas)

Anabasis (a-nab'-ạ-sis)

anabolism (a-nab'-ô-lizm)

Anabrus (an-ab'-rus)

Anacharis* (an-ak'-âr-is)

anadromous (an-ad'-rô-mus)

anaemic, anemic (a-nē'-mik; a-nem'-ik)

Anaeretes (a-nēr-ē'-tēz)

anaërobe (an-ā'-ê-rōb)

anaërobic (an-ā-ê-ro'-bik, an-ā-er'-ob-ik)

Anastrepha* (an-as'-tref-ạ)

Anagallis* (an-a-gal'-is)

anaides (a-nā-ī'-dēz) reckless, shameless.

analogous (a-nal'-ô-gus)

analogy (a-nal'ô-ji)

anamnia (an-am'-ni-ạ)

Ananas* (a-nä'-nas)
Anaphalis (a-naf'-al-is)
Anaphora (an-af'-ôr-a)
anapleurite (an-a-plŏŏr'-īt)
Anas (ā'-nas)
Anasa (ān'-as-a)
anastasis (an-as-tā'-sis) erection.
anastomosis (a-nas-tô-mō'-sis)
Anatis (an-ā'-tis)
anatomy (a-na'-tô-mi)
anatonus (a-na'-to-nus) extending upward.
anatropous (a-nat'-rô-pus)
anax (an'-aks) a lord.
Anchitherium (ang-ki-thē'-ri-um)
Anchusa* (ang-kū'-sa)
ancipital (an-sip'-it-al)
Ancistrocladus* (an-sis-trok'-lad-us)
Ancistrodon (an-sis'-trô-don)
ancylus (an'-si-lus) bent, crooked.
Andira* (an-dī'-ra)
Andrena (an-drē'-na)
Andrenidae (an-dren'-i-dē)
andricolus (an-drik'-ol-us) man-dwelling.
Andricus (an'-dri-kus
androecium (an-drē'-shi-um)
Andrographis* (an-drog'-rà-fis)
androgynal (an-droj'-i-nal)
Androloma (an-drol-ō'-ma)
Andromeda* (an-drom'-ê-da)
Andropadus (an-drop'-ad-us)
androphorous (an-drof'-ôr-us)
Andropogon* (an-drô-pō'-gōn, an-drop-ō'-gōn)

Androsace* (an-dros'-as-ē)
andrus (an'-drus) with stamens.
Aneides (an-ī'-dēz)
anemic (an-ē'-mik)
Anemone* (a̯-nem'-ō̯-nē)
Anemonella (a̯-nem-ō̯-nel'-a̯)
anemotaxis (a-nem-ō̯-taks'-is)
Angelica* (an-jel'-i-ka̯)
angiocarpus (an-ji-ō̯-kâr'-pus) vessel-fruited.
angiostomatous (an-ji-ō̯-stom'-at-us)
angiotonin (anj-i-ot'-on-in)
anglicus (ang'-li-kus) English.
angora (an-gō'-ra̯)
Anguidae (ang'-gwi-dē)
Anguilla (an-gwil'-a̯)
Anguis (an'-gwis)
angularis (ang-ŭl-ār'-is) having angles or corners.
angustifrons (an-gust'-i-fronz) narrow forehead.
angustifolius (an-gust-i-fol'-i-us, an-gust-i-fō'-li-us) with narrow leaves.
angustissimus (an-gust-is'-im-us) most narrow.
Anhima (a̯-nyē'-ma̯)
Anhinga (an-hing'-a̯, a̯-nying'-a̯)
ani (ä'-nē)
animosus (an-i-mō'-sus) bold, spirited.
Anisomeles* (an-is-om'-e-lēz)
anisatum (an-īs-ā'-tum) anise-scented.
Anisocoma* (a-nis-ō'-kom-a̯)
anisodorus (an-is-ō̯-dō'-rus) anise-odored.
Anisolabis (an-is-ol'-ab-is)
anisophyllus (an-īs-of-il'-us) unequal-leaved.
Anisota (an-is-ōt'-a̯)

Ankylosaurus (ang-ki-lȏ-sô′-rus)
anlage (än′-läg-e; pl. än′-läg-en)
annalis (an-ā′-lis) annual, continuing a year.
annectens (an-ek′-tenz) joining, connecting.
Annelida (ȧ-nel′-id-ạ)
Annonaceae* (an-ȏ-nā′-sȇ-ē)
annosus (an-ō′-sus) aged, old.
annotinus (an-ō′-ti-nus) a year old.
annulipes (an-ul′-i-pēz) ringed-foot.
annulose (an′-ȗ-lōs)
annuus (an′-u-us) yearly, annual.
Anoa (a-nō′-ạ)
Anobiidae (an-o-bī′-i-dē)
Anoda* (a-nō′-dạ)
Anolis (ȧ-nō′-lis)
anomalus (ȧ-nō′-ma-lus) irregular, deviating from rules.
anomocerus (an-o-mo′-se-rus) irregular or unequal-horned.
Anona* (ȧ-nō′-nạ)
anopetalus (an-op-et′-al-us) erect-petaled.
Anopheles (an-of′-ȇ-lēz)
Anoplophrya (an-op-lȏ-frī′-ạ)
Anoplura (an-o-plū′-rạ)
Anosia (ȧ-nō′-shi-ạ, ȧ-nō′-si-ạ)
anostraca (an-os′-trȧ-kạ) a shell.
Anous (an′-ȏ-us)
ansatus (an-sā′-tus) having a handle.
Anser (an′-sêr)
Anseres (an′-ser-ēz)
anserinus (an-se-rī′-nus) pertaining to geese.
antebrachium (an-te-brā′-ki-um)

Antechinomys (an-te-kī'-nȯ-mis)
Antechinus (an-tek-īn'-us)
Antennaria* (an-te-nā'-ri-ạ)
Anteon (an-tē'-on)
Anthaenantia*(an-thē-nan'-shi-ạ, an-thē-nan'-ti-ạ)
Anthemis* (anth'-e-mis)
Anthericum* (an-ther'-ik-um)
anthesis (an-thē'-sis)
Anthocerotae (an-tho-ser-ō'-tē)
Anthochloa* (an-thȯ-klō'-ạ)
Anthocoridae (an-thō-kȯr'-i-dē)
Anthomyiidae (an-thō-mī-ī'-i-dē)
Anthonomus (an-thon'-om-us)
Anthophoridae (an-thȯ-fȯr'-i-dē)
Anthoxanthum* (an-thȯ-zan'-thum)
anthracinus (an-thra'-sin-us) coal-colored.
Anthrenus (an-thrē'-nus)
Anthribus* (an'-thri-bus)
Anthriscus* (an-thris'-kus)
anthropeic (an-thrȯ-pē'-ik)
anthropoid (an'-thrȯ-poyd)
anthropometry (an-thrȯ-pom'-et-ri)
anthropomorphosis (an-thrȯ-pȯ-môr-fō'-sis)
anthropophora (an-thrȯ-pof'-ȯ-rạ) man-bearing.
Anthurium* (an-thū'-ri-um)
Anthus (an'-thus)
Anthyllis* (an-thil'-is)
anticus (an-tī'-kus) foremost.
Antidorcas (an-ti-dȯr'-kas)
antidromic (an-tid'-rȯ-mik)
Antigonon* (an-tig'-ȯ-non)
Antilocapra (an-til-ȯ-kap'-rạ, an-ti-lȯ-kā'-prạ)

antimeres (an'-ti-mērz)
Antiopa (an-ti'-op-ạ) wife of Lycus, king of Thebes.
Antiopella (an-ti-op-el'-ạ)
Antipatharia (an-ti-pa-thā'-ri-ạ)
antipodal (an-tip'-ŏ-dal)
antiquus (an-ti'-kwu-us) antiquated, old.
Antirrhinum* (an-ti-rī'-num)
Antispila (an-tis'-pi-lạ)
Antrostomus (an-tros'-tŏ-mus)
anus (ā'-nus)
Anychia* (a-nik'-i-ạ)
aonyx (a-on'-iks)
aorta (ā-ôr'-tạ)
Aotes (ā-ō'-tēz)
aoudad (ā'-ŏ-dad)
apache (ạ-pach'-ē)
Apachyidae (ā-pak-i'-i-dē)
Apaganthus* (a-pag-an'-thus)
Apanteles (a-pan'-te-lēz)
Apargia* (ap-âr'-ji-ạ)
Apargidium* (a-pâr-jid'-i-um)
apatelius (ap-at-el'-i-us) deceitful, deceptive.
Apatelodes (ap-at-el-ōd'-ēz)
Apaturia* (ap-at-ū'-ri-ạ)
Apera* (a'-per-ạ)
apertus (a-per'-tus) open, free.
apetalus (ap-et'-al-us) without petals.
apex (ā'-peks, pl. ap'-i-sēz or ā'-pi-sēz)
Aphanostephus* (af-a-nos'-tef-us)
Aphanostoma (af-an-ŏ-stō'-ma, af-ạ-no-stō'-mạ)
Aphelandra* (af-el-an'-drạ)
Aphelinidae (af-el-in'-id-ē)

Aphelinus (af-el-ī'-nus)
Aphelocoma (af-e-lo'-ko-mạ)
Aphelopus (af-e'-lo-pus)
Aphididae (ạ-fid'-i-dē)
aphis (ā'-fis, pl. ā'-fid-ēz)
aphlebia (af-leb'-i-ạ)
aphodal (af'-ŏd-al)
Aphodiidae (af-ŏ-dī'-i-dē)
Aphodius (af-ōd'-i-us)
Aphorista (af-ôr-ist'-ạ)
aphorodemus (ạ-fôr-od'-em-us) not bearing a body.
Aphriza (af'-ri-zạ)
Aphrophora (af-rof'-ôr-ạ)
Aphyllon* (a'-fil-on)
aphyllus (ạ-fil'-us) leafless.
apiary (ā'-pi-er-i)
apical (a'-pik-al; ā'-pik-al)
apicalis (a-pik-ā'-lis) concerning or of the top.
apiculture (ap-i-kul'-tŭr)
Apidae (ap'-i-dē)
apiferus (ap-if'-er-us) bee-bearing.
Apios* (ap'-i-os)
Apistes (ap-is'-tēz)
Apium* (ap'-i-um, ā'-pi-um)
Aplectrum* (ạ-plek'-trum)
Aplodontia (ap-lŏ-don'-shi-ạ)
Aplopappus* (ap-lŏ-pap'-us)
aplostemonous (ap-lŏ-stem'-on-us)
Aplysia (ap-lis'-i-ạ)
Apochrysidae (ap-ŏ-kris'-i-dē)
Apocynum* (ạ-pos'-i-num)
Apoda (ap'-ŏ-dạ)

apodeme (a′-pŏ-dēm)
Apodemus (a-pod′-e-mus)
Apodes (ap′-ŏ-dēz)
apodus (ap′-od-us) without feet.
Aponogeton* (a-pon-ŏ-jē′-ton)
apophysis (ap-of′-i-sis, pl. ap-of′-is-ēz)
aporogamy (à-pôr-og′-am-i)
aporrhysa (ap-or′-is-ạ)
aposematic (ap-os-em-at′-ic)
aposporogony (ap-ŏ-spôr-og′-ŏn-i)
Apostraphia (ap-os-traf′-i-ạ)
appendage (ap-en′-dej, ap-en′-dāj)
applanatus (ap-lan-ā′-tus) to, toward, flattened.
appositus (ap-oz′-i-tus) placed near, added to.
appropinquatus (ap-rop-in-kwā′-tus) drawing near.
apricot (ā′-pri-kot, ap′-ri-kot)
apricus (à-prī′-kus) lying open, exposed; also, coming from the south.
Aptenia* (ap-tē′-ni-ạ)
Aptenodytes (ap-ten-ŏ-dī′-tēz)
Aptera (ap′-têr-ạ)
Apus (ā′-pus)
aquarium (ak-wā′-ri-um)
aquatic (à-kwat′-ik, a-kwot′-ik)
aquaticus (à-kwā′-ti-kus) growing in or near water.
aquatilis (à-kwā′-til-is) living or growing in or near water.
Aquifolium* (ak-wi-fol′-i-um, ak-wi-fō′-li-um)
Aquila (ak′-wi-lạ)
Aquilegia* (ak-wi-lē′-ji-ạ)
aquiline (ak′-wi-lin)
aquilinus (ak-wil-ī′-nus) pertaining to an eagle.

Ara (ä′-rä, ā′-rạ)
Arabidopsis* (ar-a-bi-dop′-sis)
Arabis* (ar′-ȧ-bis)

Arabis. New Latin <Gr. *Arabis*, Arabian. Accent on first syllable which contains a short *a*. Pronounced: ar′-a-bis.

Arachis* (ar′-ȧ-kis)
arachnites (a-rak-nīt′-ēz) spider-like.
Aradidae (a-rad′-i-dē)
Aradus (âr′-ad-us)
aralensis (ar-al-en′-sis) from the Aral Sea.
Aralia* (ȧ-rā′-li-ạ)
Aramus (ar′-a-mus)
Aranea (ȧ-rā′-nê-ạ)
araneus (a-rā′-ne-us) pertaining to a spider.
araniferus (ȧ-rā-ni′-fer-us) spider-bearing.
Araucaria* (ar-ô-kā′-ri-ạ)
Arbacia (âr-ba′-shi-ạ)
arboreal (âr-bō′-rê-al)
arboretum (âr-bôr-ē′-tum)
arboreus (âr-bō′-re-us) tree-like.
arbuscula (âr-bus′-ku-lạ) a little tree.
Arbutus* (âr′-bŭ-tus)
Arcella (âr-sel′-ạ)

Arceuthobium* (âr-sů-thō′-bi-um)
Archaeopteryx (âr-kê-op′-têr-iks)
archegonium (âr-kê-gō′-ni-um)
archenteron (ârk-en′-têr-on)
archespore (âr′-ke-spōr)
archetypal (âr′-kê-tīp-al)
Archilochus (âr-ki-lō′-kus)
Archippus (âr-kip′-us) name of a Greek poet.
Archytas (âr′-ki-tas)
Arctictis (ârk-tik′-tis)
arcticus (ârk′-ti-kus) arctic, northern.
Arctiidae (ârk-tī′-i-dē)
Arctium* (ârk′-shi-um; ârk′-ti-um)
Arctomys (ârk′-tŏ-mis)
Arctostaphylos* (ârk-tŏ-staf′-i-los)
arcuatus (âr-ku-ā′-tus) bent, curved.
arcularius (âr-ku-lā′-ri-us) pertaining to or of a box.
arculus (âr′-ku-lus)
Ardea (âr′-dê-ạ)
ardens (âr′-denz) glowing, fiery.
Ardetta (âr-det′-ạ)
Areca (ar′-ê-kạ, a-rē′-kạ)
Arecastrum* (âr-e-kas′-trum)
arefactus (ā-re-fak′-tus) dried up, broken down, withered.
Arenaria* (a-rē-nā′-ri-ạ)
arenarius (a-rē-nā′-ri-us) of or pertaining to sand.
Arenicola (âr-ên-ik′-ŏl-ạ)
arenicolor (âr-ēn-i′-kul-ôr) sand+color.
areniferus (âr-ēn-if′-er-us) sand-bearing.
Arenivaga (âr-ēn-i-vā′-gạ)
arenosus (âr-ē-nō′-sus)

areola (ar-ē'-ŏ-lạ)
areolar (ar-rē'-ŏ-lâr)
Arethusa* (ar-e-thū'-sạ)
argali (âr'-ga-li) Mongolian word for sheep.
Argemone* (âr-je-mō'-nē)
argenteus (âr-jen'-te-us) silvery-white.
Argidae (âr'-ji-dē)
argillaceus (âr-jil-ā'-se-us) of clay, clay-colored.
Argiope (âr-jī'-ŏ-pē)
Argusianus (âr-gus-i-ā'-nus)
argutus (âr-gū'-tus) bright, lively, noisy, rattling;
  also, sharp, pungent, sly, etc.
Argynnis* (âr-ji'-nis)
argyreus (âr-ji'-re-us) silvery.
argyroneurus (âr-jir-on-ū'-rus) silver-nerved or
  -threaded.
Argyropa (ar-ji-rō'-pạ)
Argyropidae (ar-ji-rop'-i-dē)
Argythamnia* (âr-ji-tham'-ni-ạ)
aridus (ar'-id-us) withered, dry.
arietinus (ar-i-et-ī'-nus) like a ram's head.
aril (ar'-il)
Arilus (ar'-il-us)
Arisaema* (ar-i-sē'-mạ)
aristatus (ar-is-tā'-tus) furnished with an awn,
  having ears of corn.
Aristida* (ar-ist'-idạ)
Aristolochia* (ar-is-tŏ-lō'-ki-ạ), ar-is-tŏ-lok'-i-ạ)
Aristonetta (ar-ist-ŏ-net'-ạ)
aristotelian (ar-ist-ot-el'-i-an)
Arixeniidae (ar-iks-en-ī'-i-dē)
Arizona (a-ri-zō'-nạ)

armenius (âr-mē'-ni-us) of Armenia.
Armeria* (âr-mē'-ri-ạ)
armigerus (âr-mi'-jer-us) armor bearing, armed.
Armoracia* (âr-mŏ-rā'-shi-ạ); ar-môr-ā'-si-ạ)
Arnica* (âr'-ni-kạ)
Arnoseris* (âr-nos'-êr-is)
aromaticus (ar-om-at'-ik-us) spicy, fragrant.
Aronicum (ar-ō-nī'-kum)
Arquetella (âr-kwe-tel'-ạ)
arrector (ar-ek'-tôr)
arrenotokous (âr-en-ot'-ŏ-kus)
Arrhenatherum* (âr-en-a'-thêr-um)
arrhizus (ar-īz'-us) without roots.
arsipus (âr'-si-pus) with elevated or raised foot.
Artabotrys* (âr-tab'-ot-ris)
Artamus (âr'-tả-mus)
Artediellus (âr-ted-i-el'-us)
Artemia (âr-tē'-mi-ạ)
Artemisia* (âr-tê-mis'-i-ạ, ar-tê-mish'-i-ạ)
Arthrodira (âr-thrŏ-dī'-rạ)
Arthrolobium* (âr-thrŏ-lō'-bi-um, ar-thrŏ-lob'-i-um)
Arthromacra (âr-thrŏ-mak'-rạ)
arthropod (âr'-thrō-pod)
Arthropoda (âr-throp'-o-dạ)
Artibeus (âr-ti'-be-us)
articulatus (âr-ti-kul-ā'-tus) jointed, furnished with joints.
artus (âr'-tus) close, confined, short, straight, narrow.
Arum* (ā'-rum)
arundinaceus (a-run-di-nā'-se-us) reed-like.

Artemisia. Named in honor of Artemis. To this name is appended the Latin or Greek -*ia*, an ending often added to commemorative names. The *i* of the antepenult is short which takes the accent. Pronounced: âr-te-mis′-i-ą.

Arundinaria* (a-run-di-nā′-ri-ą)
Arundo* (a-run′-do)
arvalis (âr-vā′-lis) of cultivated field, growing on arable land.
Arvicanthis (âr-vi-kan′-this)
Arvicola (âr-vik′-ȯ-lą)
arytenoid (âr-i-te′-noyd, ȧ-rit′-ė-noyd
Asaphes (as′-ȧ-fēz)
Asarum* (as′-âr-um)
Ascalaphidae (as-ka-laf′-i-dē)
Ascaphus (as-kā′-fus)
Ascaridae (as-ka′-ri-dē)
Ascaris (as′-kȧ-ris)
Ascetta (a-set′-ą)
Aschelminthes (ask-hel-min′-thēz)
ascidium (ȧ-sid′-i-um)
Asclepias* (as-klē′-pi-as)
Asclepiodora* (as-klē-pi-ȯ-dō′-rą)
Asclera (as′-kle-rą)
ascogenous (as-koj′-en-us)

Ascomycetes* (as-kṏ-mī-sē'-tēz)
ascyphous (as'-i-fus)
Ascyrum* (a-sī'-rum)
asellus (as-el'-us) a small ass.
asemus (a'-se-mus) without a flag, i.e., without
   distinguishing mark.
asepsis (a-sep'-sis)
asexual (a-seks'-ū-al)
asilid (a-sīl'-id)
Asilidae (a-sil'-i-dē)
Asilus (a-sīl'-us)
Asimina* (a-sim'-i-nạ)
asininus (a-sin-ī'-nus) of or produced by an ass;
   also foolish.
asio (a'-si-ō) a horned owl.
asomatus (as-ōm'-at-us) incorporeal.
asparagoides (as-par-ag-o-ī'-dēz) asparagus-like.
Asparagus* (as-par'-a-gus)
Aspasia (as-pā'-shi-ạ) Aspasia, friend of Socrates.
aspera (a'-spêr-ạ) rough, uneven, fem. of *asper*.
asperatus (as-pêr-ā'-tus) made rough, uneven;
   also, exasperating.
Aspergillus* (as-pêr-jil'-us)
aspernatus (as-pêr-nā'-tus) despised, rejected.
aspersus (a-spêr'-sus) scattered, sprinkled over
   with, bespattered.
Asperugo* (as-pḛ-roo'-gō)
Asperula* (as-pêr'-ûl-ạ)
Asphodelus* (as-fod'-el-us)
Aspidistra* (as-pid-is'-trạ)
Aspidium (as-pid'-i-um)
Aspidonectes (as-pi-dṏ-nek'-tēz)

Aspila (as'-pi-lạ)

Asplenium (as-plē'-ni-um)

assessor (as-es'-ôr) an aide, he that sits by one.

assimilation (a-sim-i-lā'-shun)

assimilis (à-sim'-il-is) similar, like.

association (à-sō-si-ā'-shun, à-sō-shi-ā'-shun)

Astacus (as'-tak-us)

Asterias (as-tē'-ri-as)

Asterophrys (as-tê-rof'-ris)

asthenic (as-then'-ik)

asthma (az'-mạ, as'-mạ)

Astilbe* (à-stil'-bē)

astomatous (as-tōm'-at-us)

astomous (as'-tôm-us)

astomus (as'-tom-us) without a mouth.

Astragalinus (as-trag-al-ī'-nus)

Astragalus* (as-trag'-a-lus)

Astrantia* (as-tran'-shi-ạ)

astreans (as'-tre-anz) star-like, gleaming like a star.

Astur (as'-têr)

Astyanax (as-tī'-à-naks)

asymmetrical (a-si-met'-ri-kal)

asyndetus (a-sin'-det-us) without connection.

Atamasco* (at-am-as'-kō)

Atamisquea* (at-am-is'-kwe-ạ)

atavic (at-av'-ik)

atavism (at'-av-izm)

atavus (at'-à-vus) an ancestor.

Ateleopus (at-e-lē'-ô-pus)

ateleosis (à-tel-ê-ō'-sis)

Atelerix (à-tel'-er-iks)

Ateles (at′-e-lēz)

ater (ā′-ter) black.

aterrimus (ā-ter′-i-mus) pronouncedly black.

Athene (ath-ē′-nē)

Atherinidae (ā-thêr-in′-i-dē)

Atherura (ath-ē-rū′-ra̧)

Athyrium* (a̧-thir′-i-um)

Athysanus* (a̧-this′-an-us)

atmosteon (at-mos′-te-on)

atokus (at′-ok-us)

atoll (a̧-tol′, at′ol)

Atragene* (a̧-traj′-ê-nē)

atratus (ā-trā′-tus) dressed in black.

atretic (a̧-trē′-tik)

atricapillus (ā-tri-kap-il′-us) black-haired, black-capped.

atricristatus (ā-tri-kris-tā′-tus)   black+combed, tufted, crested.

Atriplex* (at′-ri-pleks)

atrium (ā′-tri-um, pl. ā′-tri-a̧) a room, a hallway.

atrofuscus (ā-trō-fus′-kus) dark-brown.

atrogularis (ā-trô-gūl-ā′-ris) with black throat.

Atropa* (at′-rop-a̧)

Atropidae (a-trop′-i-dē)

atrorubens (ā-trō-rub′-enz) black, dark red.

atrous (ā′-trus)

atrovirens (ā-trō′-vir-enz) blackish-green.

atrox (ā′-troks) fierce, horrible, dark, gloomy.

Attagenus (at-a̧-jēn′-us)

Atypinae (at-i-pī′-nē)

Atypus (at′-i-pus)

Aucuba* (ô-kū′-ba̧)

auchenium (ô-kē′-ni-um)
aucuparius (ô-kup-ā′-ri-us) watched for.
audax (ô′-daks) spirited, audacious.
augescens (ô-jes′-enz) increasing, multiplying.
Aulostoma (ô-los′-tô-mą)
aurantiacus (ô-ran-tī′-ak-us) orange-colored.
auratus (ôr-ā′-tus) gilded, covered with gold.
Aurelia (ô-rē′-li-ą)

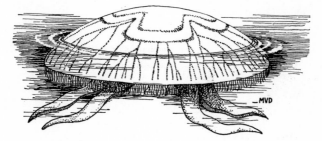

Aurelia <L. *Aurelia* (lit. *golden*), a feminine name. The accent falls on the antepenult which contains a long *e*. Pronounced: ô-rē′-li-ą.

aureus (ô′-re-us) of gold, golden.
auricestus (ôr-i-kes′-tus) with golden girdle.
auricomus (ô-rik′-om-us) with golden hair, with golden foliage.
auricula (ô-rik′-û-lą)
Auriparus (ô-ri′-pa-rus)
auritus (ô-rī′-tus) with ears, having large ears.
aurochs (ô′-roks, ou′-roks)
austerus (ôs-tē′-rus) harsh, tart; also, severe, rigid, stern, troublesome.
australis (ôs-strā′-lis) southern.
austriacus (ôs-tri′-ak-us) belonging to the south.

austrinus (ô-strī'-nus) southern.
autochthon (ô-tok'-thon)
autochthonous (ô-tok'-thŏn-us)
Autodetus (ô-tod'-ê-tus)
autoecious (ô-tē'-shi-us)
autolysis (ô-tol'-is-is)
Autolytus (ô-tol'-i-tus)

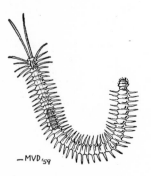

Autolytus <Gr. *auto-* <*autos* self+*lytos*,
dissolvable. A marine annelid. Pronounced: ô-tol'-i-tus, not ô-tō-lī'-tus.

— MVD '59

Automeris (ôt-om'-er-is)
autosome (ô'-tŏ-sōm)
autotomy (ô-tot'-ŏ-mi)
autotrophic (ô-tŏ-trof'-ik)
autotropic (ô-tŏ-trop'-ik)
autumnalis (ô-tum-nā'-lis) belonging to autumn.
auxin (ôk'sin)
avarus (av-ā'-rus) greedy.
Avena* (ȧv-ē'-nạ)
avenaceus (av-ē-nā'-se-us) of oats.
aveniform (ȧv-ē'-ni-fôrm) having the form of oats.
Averrhoa* (av-e-rō'-ạ)
aversus (a-vêr'-sus) turned away, bent back.

Aves (ā′-vēz)

Avicennia* (av-i-sen′-i-ạ)

avicularis (av-ik-ul-ā′-ris) pertaining to little birds.

avitus (av-ī′-tus) ancestral.

avium (a′-vi-um) a desert, a wilderness.

avius (ā′-vi-us) remote, solitary, lonely.

avocet (av′-ô-set)

axcipetal (aks-ip′et-al)

axilla (ak-sil′-ạ, ak′-sil-ạ)

axillaris (ak-sil-ā′-ris) borne in axils, axillary.

axis (ak′-sis, pl. ak′-sēz)

axon (aks′-ōn)

Axonopus* (aks-on′-ô-pus)

Ayenia* (à-yē′-ni-ạ)

Azalea* (à-za′-le-ạ, à-zā′-lê-ạ)

Azolla* (a-zol′-ạ)

azureus (az-ū′-re-us) azure-blue.

azygoid (az′-i-goyd)

azygous (az′-i-gus)

B

Babiana* (ba-bi-ā′-nạ)

Babirussa (bab-i-rū′-sạ)

baccans (bak′-anz) with berries, berry-like, pulpy.

baccatus (bak-ā′-tus) berried.

Baccharis* (bak′-ạ-ris)

baccifer (bak′-sif-êr) berry bearing.

bacciferous (bak-sif′-êr-us)

bacciform (bak′-si-fôrm)

bacillary (ba-sil′-ar-i)

bacilliparous (bas-il-ip′-ar-us)

bacterium (bak-tēr′-i-um)
Bactrocerus (bak-trô′-se-rus)
bacula (bā′-kul-ạ) a small berry.
baculiferus (ba-kul-i′-fêr-us) bearing rods or reeds.
baculum (bak′-ul-um) a stick, staff, rod, support.
Bacunculidae (bak-un-kū′-li-dē)
badius (bad′-i-us) brown, chestnut-brown.
Baeocera (bē-os′-er-ạ)
Baeolophorus (bē-ol-of′-ôr-us)
Baetidae (bē′-ti-dē)
Bahia (bä-ē′-ạ)
Baiomys (bī′-ô-mis)
Balaeniceps (bal-ē′-ni-seps)
Balaenoptera (bal-ê-nop′-têr-ạ)
Balaninus (bal-an-ī′-nus)
Balanosphyra (bal-a-nô-sfī′-rạ)
Balanus (bal′-a-nus)
Balearica (bal-ê-ar′-ik-ạ)
baliolus (bal-i′-o-lus) dark, swarthy, chestnut-
   colored.
Balistes (bal-is′-tēz)
Ballota* (bal-ō′-tạ)
balsam (bôl′-sam)
balsameus (bal-sam′-e-us) having the soothing
   qualities of balm.
balsamiferus (bal-sam-if′-er-us) balsam-bearing.
balsamifluus (bal-sam-if′-lu-us) balsam-yielding.
Bambusa (bam-bū′-sạ)
Bambusicola (bam-bū-sik′-ô-lạ)
bambusoides (bam-bū-so-ī′-dēz) bamboo-like.
Bandicota (ban-di-kō′-tạ)
Barbarea* (bâr-bâr-ē′-ạ)

barbarus (bâr'-bâr-us) foreign.
barbatus (bâr-bā'-tus) bearded.
barbigerus (bâr-bi'-je-rus) having a beard.
Barosaurus (bâr-ȯ-sô'-rus)
basalis (bā'-sal-is) basal.
Basella* (bas-el'-ạ)
Basidiomycetes (bas-id-i-ō-mī-sē'-tēz)
basifixed (bā'-sif-iksd)
basifugal (bā-sif'-ū-gal)
basilaris (bas-il-ā'-ris) pertaining to the base.
Basileuterus (bas-il-ū'-têr-us)
Basiliscus (bas-il-isk'-us)
Basilona (bas-il-ōn'-ạ)
bassalia (bas-ā'-li-ạ)
Bassaricyon (bas-ȧ-ri'-si-on)
Bassariscus (bas-ȧ-ris'-kus)
Bathyergus (bath-i-êr'-gus)
Batis* (bā'-tis)
Batrachoseps (bat-rȧ'-kȯ-sēps)
batrachostomus (bat-rȧ-kos'-tȯ-mus)
Batrachus (bat'-rȧ-kus)
Batrisodes (bat-ris-ōd'-ēz)
Bdellostoma (del-os'-tȯ-mạ)
Bdeloura (de-lū'-rạ)
Bdeogale (de-ȯg'-a-lē)
Begonia* (bė-gō'-ni-ạ)
Belamcanda* (bel-am-kan'-dạ)
Belemnite (bel'-em-nīt)
bellicosus (bel-i-kō'-sus) full of fight.
Bellis* (bel'-is)
Bellophis (bel'-of-is)
Beloperone* (bel-ȯ-per'-on-ē, bel-ȯ-per-ō'-nē)

Belostomatidae (bel-os-tô-mat′-i-dē)
Belyla (bel-īl′-ạ)
Bembicidae (bem-bis′-i-dē)
Bembidium (bem-bid′-i-um)
Berberis* (bêr′-be-ris, bêr′-bêr-is)
Bernicla (bêr′-ni-klạ)
Beroë (ber′-ô-ē)
Berothidae (ber-ōth′-i-dē)
Berteroa* (bêr-têr-ō′-ạ)
Berula* (ber′-u-lạ)
Beryx (ber′-iks)
Bessera* (bes′-êr-ạ)
Beta* (bē′-tạ)
betae (bē′-tē) of the beet.
Bethylidae (beth-il′-i-dē)
Betonica* (bet-on′-i-kạ)
Bettongia (bet-on′-ji-ạ)
Betula* (bet′-ŭ-lạ)
Bibio (bib′-i-ō)
Bibos (bī′-bos)
bicarinatus (bik-ar-in-ā′-tus) with two keels.
bicipital (bis-ip′-it-ạl)
bicodulus (bik-ōd′-ul-us) with two tails.
bicolor (bik′-ol-ôr) two-colored.
bicornis (bik-ôr′-nis) two-horned.
bicors (bik′-ôrz) having two hearts, false.
bicrenatus (bik-rē-nā′-tus) twice scalloped.
bicruris (bik′-rûr-is) with two limbs.
Bidens* (bī′-denz, bid′-enz)
biennis (bi-en′-is) lasting two years.
bifarius (bif-âr′-i-us) double, in two ways.
bifid (bif′-id, bī′-fid)

bifidus (bif'-id-us) divided into two parts, cleft.
biflorus (bif-lō'-rus) two-flowered.
biforous (bif'-ôr-us)
bifrons (bif'-ronz) two-fronded.
bigemminate (bī-jem'-i-nāt)
Bignonia* (big-nō'-ni-ạ)
bijugate (bī'-jū-gāt)
bijugus (bij'-ug-us)
bilineatus (bil-in-e-ā'-tus) two-lined.
bilituratus (bil-it-ur-ā'-tus) twice blotted out.
bilocularis (bil-ok-ŭ-lā'-ris) with two compartments.
bimerus (bim'-er-us)
bimus (bī'-mus) lasting two years.
binaevatus (bin-ē-vā'-tus) two-spotted.
binarius (bī-nā'-ri-us)
binocular (bin-ok'-ŭ-lêr; bĭ-nok'-ŭ-lêr)
binoculatus (bin-ok-ul-ā'-tus)
binominal (bī-nom'-i-nal)
biota (bī-ō'-tạ; bī'-ot-ạ)
biotonus (bī-ot'-on-us)
biparous (bip'-ar-us)
bipedal (bī-pe'-dal; bip'-ed-al)
bipes (bi'-pēz) two-footed.
bipinnaria (bī-pi-nā'-ri-ạ)
bipunctatus (bip-unk-tā'-tus) two-spotted.
biramose (bī-rām'-ōs)
bisnaga (bis-nä'-gạ)
bisulcus (bis-ul'-kus) twice-parted.
bitegmous (bī-teg'-mus)
Bitis (bī'-tis)
Bittacidae (bi-tas'-i-dē)

Bittium (bit'-i-um)
bivalent (bī-vā'-lent, biv'-ạ-lent)
bivittatus (biv-it-ā'-tus) twice bound with a head-band.
Bixa* (biks'-ạ)
blandus (blan'-dus) smooth, agreeable, pleasant.
Blapstinus (blap'-sti-nus)
Blarina (bla-rī'-nạ)
blastema (blast'-em-ạ; blast-ēm'-ạ)
Blastocerus (blas-tôs'-er-us)
blastomere (blas'-tô-mēr)
blastula (blas'-tū-lạ, pl. blas'-tū-lē)
Blattaria (blat-ār'-i-ạ)
Blattidae (blat'-i-dē)
Blechnum* (blek'-num)
Bleo* (blē'-ō)
Blepharidachne* (blef-ar-i-dak'-nē)
blepharoplast (blef'-ar-ô-plast, blef-âr'-ô-plast)
Blephila* (blef-il'-i-ạ)
Blighia* (blī'-i-ạ)
Blissus (blis'-us)
Blitum* (blī'-tum)
Blumeanum* (blū-me-ā'-num)
Blysmus* (bliz'-mus)
Bochus (bok'-us)
Boehmeria* (bô-mē'-ri-ạ)
Boerhaavia* (bŏŏr-hä'-vi-ạ)
Bolboxalis* (bol-bok'-sȧ-lis)
Boletus* (bô-lē'-tus)
Bomarea* (bom-ā'-re-ạ, bō-mā'-re-ạ)
Bombinator (bom-bi-nā'-tôr)
Bombycidae (bom-bis'-i-dē)

Bombycilla (bom-bi-sil'-ạ)
Bombyliidae (bom-bi-lī'-i-dē)
Bonasa (bon-ā'-sạ)
bonasus (bon-ā'-sus) a buffalo.
Boopiidae (bô-op-ī'-i-dē)
Borago* (bô-rā'-gō)
borealis (bôr-e-ā'-lis) of the north.
Boreidae (bô-rē'-i-dē)
Boreomyia (bôr-ê-ô-mī'-i-ạ)
Boriomyia (bôr-i-ô-mī'-i-ạ)
Boromys (bô'-rô-mis)
Borus (bōr'-us)
Boselaphus (bos-el'-ā-fus)
Bostrichidae (bos-trik'-i-dē)
Bostrichus* (bos'-trik-us)
botanodes (bot-an-ō'-dēz) herbaceous.
Botaurus (bô-tôr'-us)
Bothrops (bō'-throps, both'-rops)
Botrychium* (bô-trik'-i-um)
Botryllus (bot-ril'-us)
botryoidal (bot-ri-ô-ī'-dal)
botrytis (bot-rī'-tis) racemose.
Botula (bot'-ŭ-lạ)
Bougainvillea* (boo-gin-vil'-lê-ạ, boo-gān-vil'-lê-ạ)
bovine (bō'-vīn; bō'-vin)
Bovista* (bô-vis'-tạ)
Boweia* (bō-wē'-i-ạ)
Boykinia* (boy-kin'-i-ạ)
bracatus (brak-āt'-us) with breeches.
brachelytra (brak-el'-i-trạ)
brachial (brā'-ki-ạl)
Brachiaria* (brāk-i-ār'-i-ạ)

brachiate (brā'-ki-āt)
brachium (brā'-ki-um, brak'-i-um)
Brachycera* (brak-i'-ser-ą)
Brachychaeta* (brak-i-kē'-tą)
Brachycome* (brak-ik'-om-ē)
brachydactyly (brak-i-dak'-til-i)
Brachyelytrum* (brak-i-el'-it-rum)
Brachylaena* (brak-il-ē'-ną)
Brachyphylla (brak-i-fil'-ą)
Brachypodium* (brak-i-po'-di-um)
brachyptera   (brak-ip'-têr-ą)   short-finned   or
    winged.
brachypus (brak'-i-pus) broad-foot.
Brachyris* (brak-ī'-ris)
Brachystola (brak-is'-tŏ-lą)
Bracon (bra'-kon)
Braconidae (bra-kon'-i-dē)
Bradypus (brad'-i-pus)
branchelion (brang-kel'-i-on)
branchia (brang-ki'-ą)
branchiopod (brang'-ki-ŏ-pod)
Branchiostoma (brang-ki-os'-tŏ-mą)
Branchipus (brang'-ki-pus)
Brandegia* (bran-dē'-ji-ą)
Brasenia* (bra-sēn'-i-ą)
Brassavola* (bras-ā'-vŏ-lą)
Brassica* (bras'-i-ką)
Braya* (brā'-yą)
bregma (breg'-mą)
Bregmaceros (breg-mas'-e-ros)
brephic (bref'-ik)
brevicalyx (brev-ik-āl'-iks) with short calyx.

brevicomis (brev-i-kō'-mis) brief or shortly oblig-
ing.
breviculus (brev-i'-ku-lus) a little short.
brevifimbriatus (brev-if-im-bri-ā'-tus) short-
fringed.
brevilabrus (brev-i'-la-brus) with short lip.
brevipes (brev'-ip-ēz) with brief or small foot.
brevis (brev'-is) short.
Breviscapa* (brev-is-kā'-pạ)
brevistylus (brev-is-tī'-lus) short-styled.
brevitubus (brev-i-tū'-bus) with short tube.
Briza* (brī'-zạ)
brizoides (brī-zo-ī'-dēz)
brochus (brok'-us) with projecting teeth.
Brodiaea* (brō-di-ē'-ạ)
Bromius (brom'-i-us)
bromoides (brōm-o-ī'-dēz)
Bromus* (brom'-us, brō'-mus)
Brongniartia* (bron-yȧr'-ti-ạ)
Brotogeris (brō-toj'-er-is)
Broussonetia* (brūs-son-ē'-ti-ạ)
Browalia* (brô-wal'-i-ạ)
Bruchus* (brū'-kus)
brumalis (brū-māl'-is) wintery, pertaining to the
shortest day.
Brunfelsia* (brun-fel'-shi-ạ)
brunneus (brun'-e-us) brown.
Bryaxis (brī-aks'-is)
Brycon (brī'-kon)
Bryonia* (brī-ō'-ni-ạ)
Bryophyta (brī-of'-itạ)
Bubalis (bū'-ba-lis)

bubalus (bū'-ba-lus) of the wild-ox.
Bubo (bū'-bō)
bucca (buk'-ạ, pl. buk'-ē)
buccal (buk'-al)
buccatus (buk-āt'-us) big-jawed, with big cheeks.
buccinatorius (buk-sin-at-ôr'-i-us) known, proclaimed.
Buccinum (buk'-si-num)
bucephalus (bū-sef'-al-us) bull-headed.
Buceros (bū'-ser-os)
Buchloe* (bŭ-klō'-ē)
Bucida* (bū'-sid-ạ)
buculus (bū'-ku-lus) a bullock.
Bucyon (bū'-si-on)
Buddleja* (bud'-lê-yạ)
Bufo (bū'-fō)
Bufonidae (bŭ-fon'-id-ē)
bufonis (bū-fō'-nis) of toads.
bufonius (bū-fōn'-i-us) having to do with toads.
Bugula (bū'-gu-lạ)
bulbifera (bulb-if'-er-ạ) bulb-bearing.
Bulgaria* (bul-gā'-ri-ạ)
Bulimus (bū'-li-mus)
bulla (bōō'-lạ)
bullatus (bul-ā'-tus) inflated.
Bumelia* (bŭ-mē'-li-ạ)
Bungarus (bung'-gȧ-rus)
Bunium* (bū'-ni-um)
bunodont (bū'-nô-dont)
bunoid (bū'-noyd)
bunophilus (bū-nô'-fil-us) hill-loving.
Buphaga (bū'-fȧ-gạ)

Buphthalmum* (būf-thal'-mum)
Bupleurum* (bū-plū'-rum)
Burhinus (bŭ-rī'-nus)
burrus (bûr'-us) red.
bursa (bûr'-sạ, pl. bûr'-sē) a pouch.
bursarius (bûr-sā'-ri-us) pouched.
Bursera* (bûr'-sêr-ạ)
bursiformis (bûr-si-for'-mis) pouch-shaped, pocket-like.
Busycon (bŭ-sī'-kon)
Buteo (bū'-tê-ō)
Buthus (bū'-thus)
Butia* (bū'-ti-ạ)
Butomus* (bū'-to-mus)
Butorides (bū-tôr-ī'-dēz)
buxifolius (buks-i-fol'-i-us, buks-i-fō'-li-us) box-leaved.
Buxus* (buk'-sus)
Bycanistes (bik-an-is'-tēz)
Byrrhus (bir'-us)
Byrsonima* (bir-son'-im-ạ)
byssus (bis'-us, pl. bis'-us-ēz)
Bystropogon* (bis-trop-ō'-gōn)
Byturus* (bit-ū'-rus)

C

caballus (ka-bal'-us) an inferior pony, a nag.
Cabomba* (kab-om'-bạ)
Cacalia* (kak-ā'-li-ạ)
Cacatua (kak-ạ-tū'-ạ)
cachinnans (kak'-in-anz) laughing.

cadaver (ka-da′-vêr, pl. ka-da′-vêr-ą; ka-dā′-ver)
cadaveric (ka-dav′-êr-ik)
caddis (kad′-is)
caducous (kad-ū′-kus)
Cadulus (kad′-ŭ-lus)
Caecidotea* (sē-si-dō-tē′-ą)
Caeciliidae (sē-si-lī′-i-dē)
caecum (sē′-kum)
caecus (sē′-kus) blind; also, hidden, obscure.
caelatus (sē-lā′-tus) carved, engraved.
caelebs (sē′-lebz) unmarried, single.
Caenidae (sē′-ni-dē)
Caenolestes (sē-nŏ-les′-tez)
caenosus (sē-nō′-sus) muddy.
caerulescens (sē-rul-es′-senz) becoming blue.
caeruleus (sē-ru′-le-us) dark-colored, dark blue or
    green, blue like the surface of the sea.
Caesalpinia* (ses-al-pin′-i-ą)
caesius (sē′-si-us) bluish-gray.
caespitosus (sē-spi-tō′-sus) tufted.
cafer (kaf′-êr) of Caffraria (Kafir).
caffer (kaf′-êr) Kafir (Kaffir), in South Africa.
Caiman (kā′-man)
Cairina (kā-rī′-ną)
Cajanus* (ka-jā′-nus)
Cakile* (ka-kī′-lē)
Caladium* (kal-ā′-di-um)
Calamagrostis* (kal-ȧ-mȧ-gros′-tis)
Calamites* (kal-am-ī′-tez)
Calamoceratidae (kal-ȧ-mō-ser-at′-i-dē)
Calamospiza (kal-a-mŏ-spīz′-ą)
Calamovilfa* (kal-a-mŏ-vil′-fą)

Calandrinia* (kal-an-dri′-ni-à)
Calanthe* (ka-lan′-thê)
calathinus (kal-ath-ī′-nus) basket-like.
calcaratus (kal-kar-ā′-tus) spurred.
calcareous (kal-kā′-rê-us, kal-kâr′-ê-us)
calcareus (kal-kā′-re-us) pertaining to lime.
Calcarius (kal-kā′-ri-us)
calefacient (kal-ê-fā′-shent)
Calendula* (kal-en′-dū-lạ)
calendulus (kal-en′-dul-us) of the first of the month.
Calendulus (kal-en′-du-lus)
Calidris (kal-id′-ris)
calidus (kal′-i-dus) warm, hot.
caligatus (kal-i-gā′-tus) booted, wearing boots.
Caligatus (kal-i-gā′-tus)
caliginosus (kā-lī-jin-ōs′-us) obscure, dark, covered with mist.
Calimeris* (kal-im′-er-is)
calines (kal′-ēnz)
Caliphruria* (kal-if-rū′-ri-ạ)
Calistemma* (kal-is-tem′-ạ)
Calla* (kal′-ạ)
Calledapteryx (kal-ed-ap′-têr-iks)
Callianassa (kal-i-an-as′-ạ)
Calliandra* (kal-i-an′-drạ)
callianthemus (kal-i-an′-the-mus) beautiful-flowered.
Callicarpa* (kal-i-kâr′-pạ)
Calligonum (kal-ig′-on-um)
Calligrapha (kal-ig′-raf-ạ)
Callimome (kal-im-ōm′-ē)

calliope (kal-ī'-ō-pē) beautiful-voiced.
Callirrhoe* (kal-ir'-ō-ē)
Callistemon* (kal-i-stē'-mon)
Callistephus* (kal-is'-te-fus)
Callitris* (kal-it'-ris, kal-ī'-tris)
Callimomidae (kal-i-mōm'-i-dē)
Callipepla (kal-i-pep'-lạ)
Calliphoridae (kal-i-fôr'-i-dē)
Callisaurus (kal-i-sôr'-us)
Callitriche* (kal-it'-ri-kē)
Callizia (kal-iz'-i-ạ)
Callosobruchus (kal-os-ō-brū'-kus)
Calluella (kal-ū-el'-ạ)
Calluna* (kal-ū'-nạ)
Calobata (kal-ob'-at-ạ)
Calocalanus (kal-ok-al'-an-us)
Calocarpon* (kal-ō-kâr'-pon)
Calochortus* (kal-ok-ôr'-tus)
Calodracon* (kal-od'-rak-on)
Caloenas (kal-ē'-nas)
calogaster (kal-ō-gas'-ter) with beautiful belly.
Caloglossa* (kal-og-los'-ạ, kal-og-lō'-sạ)
Calonectris (kal-on-ēk'-tris)
Calonyction* (kal-ō-nik'-ti-on)
Calophaca* (kal-of'-ak-ạ)
Calophanes* (kal-of'-an-ēz)
Calopogon* (kal-ō-pō'-gōn)
Caloptenus (kal-op-tē'-nus)
Calopteron (kal-op'-têr-on)
Calosoma (kal-ō-sō'-mạ)
Calotes (kal'-ō-tēz)
Calothorax (kal-oth'-ôr-aks)

Caltha* (kal'-tha)

caltrop (kal'-trop)

calycanthus (kal-ik-an'-thus) calyx-flowered.

calycine (kal'-i-sīn)

calycinus (kal-is'-in-us, kal-is-īn'-us) with persistent calyx.

Calycocarpum* (kal-ik-ȯ-kâr'-pum)

Calycodenia* (kal-ik-ȯ-den'-i-a)

Calycoseris* (kal-ik-ō'-ser-is)

Calycotome* (kal-ik-ot'-om-ē, kal-ik-ot-ō'-mē)

calyculatus (kal-ik-ul-ā'-tus) provided with a calyx.

Calydermos (kal-id-êr'-mos)

Calypso* (kal-ip'-sō)

Calypte (kal-ip'-tē)

Calyptomerus (kal-ip-tȯ-mē'-rus)

calyptraeus (kal-ip-trē'-us) hooded, helmeted.

Calystegia* (kal-is-te'-ji-a, kal-is-tē'-ji-a)

Calythrix* (kal-ith'-rix)

calyx (kā'-liks, pl. kā'-li-sēz)

Cambarus (kam'-ba-rus)

Cambrian (kam'-bri-an)

cambricus (kam'-bri-kus) of Wales (Cambria).

Camelina* (kam'-ê-lī'-na, ka-mel'-i-na)

Camellia* (ka-mel'-i-a)

Camelus (ka-mē'-lus)

campaneus (kam-pā'-ne-us) of the field.

Campanula* (kam-pan'-ū-la)

Campephilus (kam-pē'-fil-us, kam-pef'-il-us)

campestris (kam-pes'-tris) pertaining to a field, even, flat.

Campodeidae (kam-po-dē'-i-dē)

campodeiform (kam-po-dē′-i-fôrm)
Camptorhynchus (kam-ptŏ-ring′-kus)
Canace (kan′-a-sē)
Canachites (kan-a-kī′-tēz)
canalis (kan-ā′-lis) a pipe, a groove.
canariensis (kan-ā-ri-en′-sis) belonging to the
    Canary Islands.
Canavalia* (kan-av-ā′-li-ạ)
Canbya* (kan′-bi-ạ)
candicans (kan′-di-kanz) white, wooly, hoary.
candidulus (kan-did′-ul-us) shining white.
candidus (kan′-did-us) pure-white, shining.
Canella* (kan-el′-ạ)
canescens (kan-es′-enz) becoming white or gray.
Canifa (kan′-if-ạ)
canine (ka-nīn′, kā′-nīn)
caninus (kan-ī′-nus) of or pertaining to a dog.
Canis (kā′-nis)
Canistrum* (kan-is′-trum)
Canna (ka′-nạ)
Cannabis* (kan′-a-bis)
Canotia* (kan-ō′-ti′-ạ))
cantabricus (kan-tab′-ri-kus) belonging to Cantab-
    ria.
cantaloupe (kan′-tȧ-lo͞op)
Cantatores (kan-tȧ-tō′-rēz)
Cantharidae (kan-thâr′-i-dē)
Cantharis (kan′-thâr-is)
Canthon (kan′-thon)
cantianus (kan-ti-ā′-nus) of Kent.
Cantua* (kan′-tŭ-ạ)
canus (kā′-nus) ash-colored.

canutus (kā-nū'-tus) gray, hoary.
Capella (kȧ-pel'-ạ)
capercaille (kap-êr-kāl'-yê)
capibara (kap-i-bä'-rạ)
capillaris (kap-il-ā'-ris) of or pertaining to the
  hair.
capillary (kap'-i-le-ri, ka-pil'-e-ri)
capillus-veneris (kap-il'-us ven'-er-is) Venus's
  hair.
capistratus (kap-is-trā'-tus) bridled.
capitatus (kap-it-ā'-tus) headed.
capitellum (kap-it-el'-um)
Capniidae (kap-nī'-i-dē)
Capparidaceae* (kap-ä-ri-dā'-sê-ē)
capreolate (kap-rē'-ȯ-lāt, kap'-rê-ȯ-lāt)
Capreolus (kap-rē'-ȯ-lus)
Capricornis (kap-ri-kôr'-nis)
Caprimulgidae (kap-ri-mul'-ji-dē)
Caprimulgus (kap-ri-mul'-gus)
Caprinus (kap'-ri-nus)
capriolatus (kap-ri-ol-ā'-tus) having tendrils.
Capromys (kap'-rȯ-mis)
Capsella* (kap-sel'-ạ)
capuchin (kap'-ṳ-chin, kap'-ṳ-shēn)
Carabidae (kar-ab'-i-dē)
Carabus (kar'-a-bus)
Caragana* (kâr-ȧ-gā'-nạ)
Caralluma* (kar-al-lū'-mạ)
carapace (kar'-a-pās)
Carcal (kâr'-kal)
carcharias (kâr-ka'-ri-as) a kind of dog-fish.
Carcinides (kâr-sin-ī'-dēz)

Carcocapsa (kâr-kô-kap′-sạ)
Cardamine* (kâr-dam-ĭ′-nē)
cardamine (kâr′-dam-īn)
cardiaca (kâr-di′-ak-ạ) to do with the heart.
cardinalis (kâr-din-ā′-lis) cardinal-red; also, chief.
Cardiospermum* (kâr-di-ô-spêr′-mum)
cardon* (kâr-dōn′)
carduaceus (kâr-dū-ā′-se-us) thistle-like, a thistle.
Carduelis (kâr-dŭ-ē′-lis)
carduifolius (kâr-du-i-fol′-i-us, kâr-du-i-fō′-li-us) with leaves like the thistle (*Carduus*).
Carduus* (kar′-du-us)
Caretta (kâr-et′-ạ)
Carex* (kā′-reks)
Cariama (kar-i-ā′-mạ)
Carica* (kā′-ri-kạ)
caricinus (kā-ri-sī′-nus) resembling Carex.
carina (ka-rī′-nạ)
carinate (kar′-i-nāt)
carinatus (kar-i-nā′-tus) keeled.
cariosus (kar-i-ō′-sus) decayed, full of holes, withered.
Carissa* (kar-is′-ạ)
Carlina* (kâr-līn′-ạ)
Carludovica* (kâr-lud-ô-vī′-kạ)
carmineus (kâr-min′-e-us) carmine.
carnerosanus (kâr-ne-rō-sān′-us) of Carneros Pass, Mexico.
carneus (kâr′-ne-us) flesh-colored.
carnicolor (kâr-nik′-ul-ôr) flesh-colored.
carnulentus (kâr-nul-en′-tus) like flesh.
carotid (kâr-ō′-tid)

Carphibis (kâr'-fi-bis)
Carphophis (kâr-phō'-fis)
carpinifolius (kâr-pi-ni-fol'-i-us, kâr-pi-ni-fō'-li-us)
  with leaves like the hornbeam.
Carpinus* (kâr-pī'-nus)
Carpobrotus* (kâr-pŏ-brō'-tus)
Carpodacus (kâr-pod-ā'-kus)
Carpodinus* (kâr-pod-ī'-nus)
Carpophilus (kâr-pof'-il-us)
Carrisa* (kâr-is'-ạ)
Carthamus* (kâr'-tha-mus)
cartilaginus (kâr-ti-laj'-in-us) like cartilage.
Cartodere (kâr-tŏ-dē'-rē)
Carum* (kā'-rum)
caruncle (kâr'-ung-kl)
carunculatus (kâr-ung-kul-ā'-tus) like a little piece
  of flesh.
Carya* (ka'-ri-ạ, kâr'-i-ạ)
caryophyllaceus (kar-i-of-il-lā'-se-us) like *Cary-*
  ophyllum.
Caryophyllum* (kar-i-of-il'-um)
Caryopteris* (kar-i-op'-têr-is)
Caryota* (kar-i-ō'-tạ)
caryotideus (kar-i-ŏ-tid'-e-us) like caryota.
casein (kā'-se-in)
Casimiroa* (kas-i-mi-rō'-ạ)
Casmerodius (kas-mer-ōd'-i-us)
caspica (kas'-pik-ạ)
Cassandra* (kả-san'-drạ)
cassia (kash'-i-a, kas'-i-ạ)
Cassidix (kas'-i-diks)
Cassiope* (kas-ī'-op-ē)

cassis (kas'-is) a helmet.
Castanea* (kas-tan'-e-ạ, kas-tā'-nê-ạ)
castaneus (kas-ta'-ne-us) chestnut-like.
Castanospermum* (kas-tan-ŏ-spêr'-mum)
Castela* (kas'-tel-ạ)
Castilleja* (kas-til-ē'-yạ)
Casuarina* (kazh-ū-ȧ-rīn'-ạ)
catadromous (kat-ad'-rŏ-mus)
Catalpa* (ka-täl'-pạ)
Catamblyrhynchus (kat-am-bli-ring'-kus)
Catananche* (kat-ȧ-nang'-kē)
cataphyllus (ka-ta-fil'-us) with down-hanging leaves.
catena (kat-ē'-nạ)
catenatus (kat-e-nā'-tus) bound with a chain, fettered.
catenifer (kat-ē'-nif-êr) carrying or bearing a chain.
Catha* (ka'-thạ)
Catharacta (kath-âr-ak'-tạ)
Catharopeza (kath-âr-ŏ-pē'-zạ)
Cathartes (kath-âr'-tēz)
catharticus (kath-âr'-ti-kus) cleansing, purifying.
Catherpes (kath-êr'-pēz)
Cathestecum (kath-e'-ste-kum)
Catocala (kat-ok-āl'-ạ, kȧ-tok'-ȧ-lạ)
catomus (kat-ō'-mus) the shoulders.
Catoptrophorus (kat-op-tro'-fôr-us)
Catorama (kat-ôr'-a-mạ)
Catostomus (ka-tos'-tŏ-mus)
Cattleya* (kat'-le-ạ)
Caucolis* (kô'-kol-is)

caudal (kô′dal)
caudatolenticular (kô-dā-tô-len-tik′-u-lâr)
Caulanthus* (kôl-an′-thus)
caulis (kô′-lis)
Caulophyllum* (kô-lô-fil′-um)
caurinus (kôr′-i-nus) of the northwest wind.
cautus (kô′-tus) to be on guard.
cavus (kav′-us) hollow.
Ceanothus* (sē-á-nō′-thus)
Cebatha* (seb′-á-thạ)
Cebrio (seb′-ri-ō)
Cebrionidae (seb-ri-on′-i-dē)
Cebus (sē′-bus)
Cecidomyiidae (ses-i-dô-mī-ī′-i-dē)
Cecrops (sē′-krops)
Cedronella* (sē-drôn-el′-ạ)
Cedrus* (sē′-drus, sed′-rus)
Ceiba* (sā-ē′-bạ, sē-ī′-bạ)
Celama (sel-ām′-ạ)
celandine (sel′-an-dīn)
Celastrus* (sē-las′-trus)
celatus (sē-lā′-tus) hidden, kept secret.
celeratus (sel-er-ā′-tus) hastened, quickened.
Celosia* (sêl-ō′-shi-ạ)
Celsia* (sel′-shi-ạ)
Celtis* (sel′-tis)
cembroides (sem-bro-ī′-dēz) like the Cembra or Swiss Stone Pine.
cement (n. sē′-ment, v. sē-ment′)
Cemophora (sē-mof′-ôr-ạ, se-mof′-ôr-ạ)
Cenchrus* (seng′-krus)
Cenozoic (sē-nô-zō′-ik, sen-ô-zō′-ik)

Centaurea* (sen-tô′-rḗ-ą, sen-tô-rē′-ą)

Centaurium* (sen-tô′-ri-um)

Centetes (sen-tē′-tēz)

centranthifolius (sen-tran-thi-fol′-i-us, sen-tran-thi-fō′-li-us) centranthus-leaved.

Centranthus* (sen-tran′-thus)

Centrocercus (sen-trŏ-sêr′-kus)

Centrophanez (sen-trof′-ȧ-nēz)

Centrosema* (sen-trŏ-sē′-mą)

centrum (sen′-trum) a sharp point, the point around which a circle is described.

Centunculus* (sen-tun′-ku-lus)

Centurio (sen-tū′-ri-ō)

Centurus (sen-tū′-rus)

Ceophloeus (sē-of-lō-ē′-us)

Cephaelis* (sef-ā-ē′-lis)

Cephalanthera* (sef-ȧ-lan-thē′-rą)

cephalic (sē-fal′-ik, sef-al′-ik)

Cephalophus (sef-al′-ŏ-fus)

cephalopod (sef′-al-ŏ-pod, sef-al′-ŏ-pod)

Cephalopoda (sef-ȧ-lop′-ŏ-dą)

cephalopodium (sef-al-ŏ-pō′-di-um)

cephalotus (sef-al-ō′-tus) with a head.

cephalula (sef-al′-ū-lą)

Cephidae (sē′-fi-dē)

Cephus (sef′-us)

Cerambycidae (ser-am-bis′-i-dē)

Ceraphron (ser′-a-fron)

cerasifer (ser-as′-i-fêr) cherry-bearing.

cerastes (sē-ras′-tēz) a horned serpent; also, horned.

Cerastium* (ser-as′-ti-um)

Cerasus* (ser'-a-sus)
Ceratophrys (ser-à-tof'-ris)
Ceratinidae (ser-a-tin'-i-dē)
ceratocarpus (ser-a-tŏ-kar'-pus) having a horny fruit.
Ceratodus (ser-at'-ŏ-dus)
Ceratonia* (ser-à-tō'-ni-ạ)
Ceratophrys (ser-at-of'-ris)
Ceratophyllum* (ser-a-tŏ-fil'-um)
Ceratophyta (ser-a-tof'-it-ạ)
Ceratopogonidae (ser-at-ŏ-pō-gōn'-i-dē)
Ceratopsia (ser-à-top'-si-ạ)
Ceratopsyllus (ser-à-top'-sil-us)
Ceratopteris* (ser-à-top'-ter-is)
Ceratotheca* (ser-at-ŏ-thē'-kạ)
ceratus (ser-āt'-us) smeared, covered.
Cerberus (sêr'-bê-rus)
cercalis (sêr-kā'-lis) tailed
cercaria (sêr-kā'-ri-ạ)
Cerceris (sêr'-sêr-is)
Cerchneis (sêrk-nē'-is)
cerciatus (sêr-si-ā'-tus) tailed, with a tail.
cercid (sêr'-sid)
Cercidiphyllum* (sêr-sid-i-fil'-um)
Cercidium* (sêr-sid'-i-um)
Cercis* (sêr'-sis)
Cercolabes (ser-kol'-à-bēz)
Cercomys (ser'-kŏ-mis)
Cercopidae (ser-kop'-i-dē)
Cercopis (ser-kō'-pis)
Cercopithecus (ser-kŏ-pi-thē'-kus)
Cercospora* (ser-kos'-pŏ-rạ)

cercus (ser'-kus)
Cercyonis (ser-sī'-on-is)
Cerdocyon (sêr-dos'-i-on)
cere (sēr)
cerebellar (ser-ē̇-bel'-êr)
cerebellum (ser-ē̇-bel'-um)
cerebrum (ser'-ēb-rum)
Ceresa (ser-ē'-sa̤)
Cereus* (sē'-re-us)
ceriferus (sē-ri'-fer-us) producing wax, having a
     waxy covering.
cernuus (ser'-nu-us) inclined, with face toward the
     earth.
ceroma (sē-rō'-ma̤)
Ceropales (sêr-op'-a̤-lēz)
Ceropegia* (sē-rop-ē'-ji-a̤)
Ceroxylon* (sē-rok'-si-lon)
certation (sêr-tā'-shun)
Certhia (sêr'-thi-a̤)
Ceruchus (sêr'-uk-us)
cerumen (sêr-ū'-men)
Cervus (sêr'-vus)
Ceryle (ser'-i-lē)
cespitose (ses'-pi-tōs)
Cestrum* (ses'-trum)
Ceterach* (set'-êr-ak)
cetolith (sē'-tṍ-lith)
Cetoniidae (sē-tṍ-nī'-i-dē)
Cetorhinus (sē-tṍ-rīn'-us)
Cetraria (sē̇-trā'-ri-a̤)
Cettia (set'-i-a̤)
Ceuthmochares (sū-thmo-kā'-rēz)

Ceuthophilus (sū-thof'-il-us)
Ceyx (sē'-iks)
Chaenactis (kēn-akt'-is)

Chaenactis. New Latin <Gr. *chainō* to gape+*aktis*, a ray, referring to the marginal flowers of one section of the genus. The accent falls on the penult because this syllable is long (the vowel *a* followed by two consonants). Pronounced: kēn-akt'-is.

Chaenomeles* (kē-nom'-e-lēz)
Chaerophon (kē'-ro-fon)
Chaerophyllum* (kē-rô-fil'-um)
Chaeropus (kē'-rô-pus)
chaeta (kē'-ta, pl. kē'-tē)
Chaetochloa* (kē-tôk'-lô-a)
Chaetognatha (kē-tog'-nath-a)
chaetosema (kē-tos-ē'-ma)
Chaetura (kē-tū'-ra)
Chalarus (kal'-âr-us)
chalaza (kal-ā'-za)
Chalcididae (kal-sid'-i-dē)
Chalcomitra (kal-kô-mī'-tra)
Chalcophora (kal-kof'-ôra)
Chalcosiidae (kal-kos-ī'-id-ē)
Chalepus (kal'-ep-us)
Chalia (kā'-li-a)

chalice (chal'-is)
chalones (ka'-lōnz)
Chama (ka'-mạ)
Chamaea (ka-mē'-ạ)
Chamaebatia* (kam-ê-bat'-i-ạ)
Chamaecyparis* (kam-ê-sip'-âr-is, kam-ê-sip'-à-ris)
Chamaedaphne* (kam-ê-daf'-nē)
Chamaedorea* (kam-ê-dō'-re-ạ)
Chamaelirium* (kam-ê-līr'-i-um)
Chamaemyiidae (kam-ê-mī-ī'-i-dē)
Chamaerops* (kȧm-ē'-rôps)
Chamaesaracha* (kam-ē-sâr'-a-kạ)
Chameleon (kȧ-mēl'-ê-on)
Chaoboridae (kā-ỏ-bôr'-i-dē)
chaparral (sha-pâr-al')
Chara (kā'-rạ)
Characeae* (kā-rā'-sê-ē)
Charadrius (ka-rad'-ri-us)
Charina (ka-rī'-nạ)
Charionetta (kâr-i-ỏ-net'-ạ)
Charitonetta (kâr-i-tỏ-net'-ạ)
Charophycophyta (kar-ỏ-fī-kof'-it-ạ)
Chasmosaurus (kas-mỏ-sô'-rus)
Chaulelasmus (kỏ-lê-las'-mus)
Chauliodes (kỏ-li-ōd'-ēz)
Chauliognathus (kỏ-li-og'-na-thus)
Cheilanthes* (kī-lan'-thēz)
cheilanthus (kī-lan'-thus) lip-flowered.
cheilocystidium (kī-lỏ-sis-tid'-i-um)
cheiragonus (kī-rag'-on-us) with angled hand.
cheiranthoides (kī-ran-tho-ī'-dēz) like *Cheiranthus*.
Cheiranthus* (kī-ran'-thus)

Chelemys (kĕl-ē'-mis, kĕl'-ĕ-mis)
chelicera (kē-li'-sêr-ạ, pl. kē-li'-sêr-ē)
Chelidonium (kel-i-don'-i-um, kel-i-dōn'-i-um)
cheliped (kē'-li-ped)
Chelonarium (kēl-on-ār'-i-um)
Chelone* (kel-ō'-nē)
Chelonia (kel-ōn'-i-ạ)
Chelonobia (kel-ŏ-nō'-bi-ạ)
Chelopus (kēl'-ŏ-pus)
Chelydra (kel-id'rạ)
Chelyosoma (kel-i-os-ōm'-ạ)
Chen (kēn, ken)

Chen <Gr. *chēn*, a goose,
properly pronounced with
the *e* long. Through long
usage the New Latin *chen*
is now considered an allow-
able pronunciation. Pro-
nounced: kēn or ken.

Chenopodium* (kē-nop-od'-i-um)
Chermidae (kêr'-mi-dē)
Chersodromus* (kêr-sod'-rom-us)
chersophyte (kêr'-sŏ-fīt)
Chersydrus (kêr-sid'-rus)
chiasma (kī-az'-mạ)
chiastic (kī-as'-tik)

Chiliandra (kil-i-an'-drą)
Chilognatha (kī-log'-na-thą)
Chilomeniscus (kī-lô-mê-nis'-kus)
Chilomonas (kī-lôm'-ô-nas)
Chilonycteris (kī-lô-nik'-têr-is)
Chilophylla (kī-lof-il'-ą)
Chilopsis* (kī-lop'-sis)
Chilostomata (kī-lô-stom'-a-tą)
chimaera (kī-mē'-rą, kīm'-ê-rą)
Chimaphila* (kī-maf'-i-lą)
Chimonanthus* (kī-mon-anth'-us)
chimpanzee (chim-pan'-zē, chim-pan-zē')
Chincha (chin'-chą)
Chiogenes* (kī-oj'-ê-nēz)
Chionactis (kī-ôn-ak'-tis)
chionanthus (kī-ôn-an'-thus) snow-flowered.
Chion (kī'-on)
Chione (kī-ōn'-ē)
Chionodoxa* (kī-ôn-ô-dok'-są)
Chirocholus (kī-rok'-ô-lus)
Chiroleptes (kī-rô-lep'-tēz)
Chiromyces* (kī-rom'-i-sēz)
Chiromys (kī'-rô-mis)
Chironomidae (kī-rô-nom'-i-dē)
Chironomus (kī-ron'-ô-mus)
Chiropotes (kī-rop'-ô-tēz)
Chirotes (kī-rō'-tēz)
chitin (kī'-tin)
Chiton (kī'-ton)
Chitonia* (kit-ō'-ni-ą)
chlamidospore (klam-id'-os-pôr, klam'-id-ô-spôr)
Chlamydomonas (klam-id-om'-ô-nas)

Chlamydosaurus (klam-id-ô-sôr'-us)
chlamydospore (klam-id'-ô-spôr, klam'-id-ô-spôr)
Chlidanthus* (klid-anth'-us)
Chlidonias (klid-ō'-ni-as)
Chloanthes* (klō-anth'-ēz)
Chloanthez (klō-an'-thēz)
Chlora* (klō'-rạ)
chloragen (klō'-rȧ-jen)
chloragocyte (klō-rag'-ô-sīt)
chloragogen (klō-ra-gō'-jen)
chloranthus (klō-ran'-thus) green-flowered, with
greenish-yellow flowers.
chlorine (klō'-rēn, klō'-rin)
Chloris (klō'-ris)
Chloroceryle (klō-rô-ser'-i-lē)
chlorocruorin (klō-rô-krū'-ôr-in)
chlorofucin (klō-rô-fū'-sin)
Chloromonadina (klō-rō-mo-na'-di-nạ)
Chloropeltina (klō-rô-pel-tī'-nạ)
Chloropeta (klō-rop'-et-ạ)
Chloropidae (klō-rop'-id-ē)
Chlorops (klō'-rops)
chloroticus (klō-rot'-i-kus) green, pale-green.
choana (kō'-an-ạ)
choanocyte (kō'-ȧ-nô-sīt)
Choeropsis (kē-rop'-sis)
Choeropus (kē'-rô-pus)
Choisya* (koys'-shi-ạ)
choledoch (kol'-ê-dok)
cholla (chō'-yạ)
Choloepus (kō-lē'-pus)
Chondestes (kon-des'-tēz)

Chondrilla* (kon-dril'-ą)
chondrioma (kon-dri-ō'-mą)
Chondrostei (kon-dros'-tē-ī)
chone (kō'-nē)
chordata (kôr-dā'-tą)
Chordeiles (kôr-dī'-lēz)
chordorhizus (kôr-dô-rī'-zus) string-rooted.
chordotonal (kôr-dô-tō'-nal)
chordus (kôr'-dus) produced late.
chore (kō'-rē)
chorea (kô-rē'-ą)
chorion (kō'-ri-on, kôr'-i-on)
Choristidae (kō-ris'-ti-dē)
Chorizanthe* (kôr-i-zan'-thē)
Chorizema* (kôr-iz'-em-ą)
Chortophaga (kôr-tof'-ag-ą)
chresard (krē-sârd')
Chroicocephalus (krō-i-kô-sef'-ȧ-lus)
chromatin (krō'-mȧ-tin)
chromatophore (krō'-mat-ô-fôr)
Chromulina (krō-mu-lī'-ną)
Chroococcus* (krō-ô-kok'-us)
chrotorrhinus (krō-tô-rī'-nus) color+nose.
Chrozophora* (krō-zof'-ô-rą)
chrysalis (kris'-al-is, pl. kris-al'-i-dēz)
Chrysanthemum* (kris-anth'-em-um)
chryseides (kris-e-ī'-dēz)
Chrysemys (kris'-e-mis)
chryseolus (kris-e'-ol-us) golden.
Chrysididae (kris-id'-i-dē)
Chrysobalanus (kris-ô-bal'-ȧ-nus)
Chrysobothris (kris-ô-bōth'-ris)

Chrysochloris (kris-ŏ-klō'-ris)
Chrysogonum (kris-ŏg'-on-um)
chrysographes (kris-ŏg'-raf-ēz) marked with gold.
chrysolepis (kris-ō'-le-pis) golden-scaled, with golden membranes.
chrysomallus (kris-om-al'-us) with golden wool.
chrysomelas (kris-om'-el-as) golden-black.
Chrysomelidae (kris-ŏ-mel'-i-dē)
chrysomphali (kris-om'-fal-ī) golden+navel.
Chrysopelea (kris-ŏp-ēl'-e-ą)
Chrysophycophyta* (kris-ŏ-fī-kof'-it-ą)
Chrysophyllum* (kris-ŏ-fil'-um)
Chrysopidae (kris-op'-i-dē)
Chrysops (kris'-ops)
Chrysoscias* (kris-ōs'-si-as)
Chrysosplenium* (kris-ŏ-splē'-ni-um)
Chrysothamnus* (kris-ŏ-tham'-nus)
Chrysothemis* (kris-oth'-em-is)
Chrysotis (kris-ō'-tis)
chrysotoxum (kris-ot-oks'-um) golden-arched.
Chrysoxylon* (kris-ŏ-zī'-lon)
Chthamalus (tham'-à-lus)
Chyliza (kī-lī'-zą)
Chyphotes (kī-fō'-tēz)
Chytraculia* (kī-trak-ū'-li-ą)
Chytrids (kī'-tri-dz, ki'-tri-dz)
cibarian (sib-ā'-ri-an)
cibarium (sib-ā'-ri-um)
cibarius (sib-ā'-ri-us) suitable for food.
ciborium (si-bō'-ri-um)
Cibotium* (sib-ō'-ti-um)
cicada (si-kā'-dą)

Cicadellidae (sik-a-del'-i-dē)
Cicadidae (si-kad'-i-dē)
cicatricial (sik-ȧ-trish'-i-al)
cicatricle (si-kat'-rikl)
cicatrix (sik'-ȧ-triks, si-kā'-triks, pl. sĭ-kȧ-trī'-sēz)
Cichladusa (sik-lad'-us-ạ)
Cichorium* (si-kō'-ri-um)
Cicindela (sis-in-dē'-lạ)
Cicindelidae (sis-in-del'-i-dē)
cicinnal (sis'-in-al)
Cicinnurus (sis-in-ūr'-us)
Cicinnus (sis-in'-us)
Ciconia (sik-ō'-ni-ạ)
Ciconiidae (sik-ọ-nī'-i-dē)
Ciconiiformes (si-kō-ni-i-fôr'-mēz)
Cicuta* (sik-ū'-tạ)
cicutarius (sik-ū-tā'-ri-us) like *Cicuta*, a genus of Umbelliferae.
ciliaris (sil-i-ā'-ris) fringed as with eye-lashes.
ciliatus (sil-i-ā'-tus) furnished with cilia or small hairs.
cilium (sil'-i-um, pl. sil'-i-ạ)
Cimbicidae (sim-bis'-i-dē)
Cimex (sī'-meks)
Cimicidae (sī-mis'-i-dē)
Cimicifuga* (sim-i-sif'- û-gạ)
cinclides (sing'-kli-dēz, pl. of cinclis.)
Cinclus (sing'-klus)
cinctipes (sink'-ti-pēz) girdle-footed.
cinctus (sink'-tus) surrounding, girdling.
Cineraria* (sin-e-rā'-ri-ạ)
cinerarius (sin-er-ā'-ri-us) pertaining to ashes.

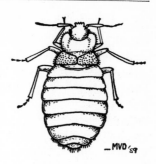

Cimex, the bed-bug. <L. *cimex*, a bug. Pronounced: sī'mex, not sim'-ex, as we often hear.

cinerascens (sin-er-as'-senz)

cinereus (sin-er'-e-us) ash-colored; like ashes.

cingulatus (sin-gul-ā'-tus) girdled, encircled, zoned.

Cinixys (sin-ik'-sis)

cinnamomeus (sin-à-mō'-me-us) of or from cinnamon.

cinnamominus (sin-à-mōm'-in-us) of or from cinnamon.

Cinnyris (sin'-i-ris)

Cinulia (sin-ū'-li-ạ)

Cionus (sī'-o-nus)

Cipura* (sip-ū'-rạ)

Circaea* (sêr-sē'-ạ)

Circaëtus (sêr-kā'-ê-tus)

Circinae (sêr-sī'-nē)

circinatus (sêr-si-nā'-tus) made round.

Circoporus (sêr-kop'-ôr-us)

circumcisus (sêr-kum-sī'-sus) cut around, cut off.

circumoesophageal (sêr-kum-ē-sof-a'-jê-al)

Circus (sêr'-kus)

cirratus (sir-ā'-tus) curled, having curls.

cirrus (sir'-us)

Cirsium* (sêr'-si-um)
Cissus* (sis'-us)
Cisticola (sis-tik'-ôl-ạ)
Cistothorus (sis-toth'-ô-rus)
Citellus (si-tel'-us)
Citheronia (sith-êr-ōn'-i-ạ)
citrinus (sit-rī'-nus) of or belonging to citrus.
citriodorus (sit-ri-ô-dō'-rus) lemon-scented.
Cixiidae (siks-ī'-i-dē)
Cladium* (klad'-i-um)
cladode (klad'-ōd)
cladogenous (klad-oj'-e-nus)
Cladoselache (klad-ô-sel'-à-kē)
Cladrastis* (klad-ras'-tis)
Clambus (klam'-bus)
clamitans (klā-mi'-tanz) loud-calling.
Clandestinus (klan-des-ti'-nus) secret, hidden.
clangula (klang'-u-lạ) a small noise.
claripennis (klà-ri-pen'-is) bright- or clear-feath-
    ered.
clarus (klā'-rus) bright, shining, evident; also, dis-
    tingushed.
clathrate (klath'-rāt, klath'-rat)
clathratus (klāth-rā'-tus) set with bars, latticed.
Clathrulina (klā-thrŭ-lī'-nạ, klath-rŭ-lī'-nạ)
clausus (klô'-sus) enclosed, shut.
Clavaria* (klàv-ā'-ri-ạ)
clavatus (klā-vā'-tus) furnished with prickles or
    points, nailed; also, furnished with stripes.
Claviceps* (klā'-vis-eps)
claviculatus (klā-vi-kŭl-ā'-tus) provided with bars,
    provided with tendrils.

clavipes (klāv′-i-pēz) club-footed.
clavus (klā′-vus)
Claytonia* (klā-tōn′-i-ạ)
cleidoic (klī-dō′-ik)
cleistogamy (klī-stog′-a-mi)
cleithrum* (klī′-thrum)
Clematis* (klē′-mat-is, klem′-à-tis)
Clemmys (klem′-is)
Cleome* (klē-ō′-mē)
Cleomella* (klē-ọ-mel′-ạ)
Cleonymus (klē-on′-i-mus)
Clepsine (klep-sī′-nē)
clepticus (klep′-ti-kus) belonging to a thief, thiev-
   ish.
Clerodendrum* (klēr-ọ-den′-drum)
cleronomy (klê-ron′-ọ-mi)
Clerus (klē′-rus)
Clethra* (klē′-thrạ, kle′-thrạ)
Clethrionomys (klē-thri-on′-ọ-mis)
climacteric (klī-mak′-ter-ik, kli-mak-ter′-ik)
Clinopodium* (klī-nop-od′-i-um)
Clinostylis (klī-nos-tī′-lis)
clitellum (klit-el′-um)
Clitoria* (klī-tô′-ri-ạ)
clitoris (klī′-tô-ris, klit′-ọ-ris)
Clivia* (klī′-vi-ạ)
cloaca (klọ-āk′-ạ)
clone (klōn)
Clonorchis (klō-nôr′-kis)
clonus (klō′-nus) confused and violent motion.
Clubionidae (klub-i-on′-i-dē)
Clupea (klū′-pê-ạ)

clusius (klū'-si-us) a cognomen of Janus.

clypeatus (klip-e-ā'-tus) shielded, with shields.

clypeus (klip'-ê-us)

clysium (kli'-si-um)

cnemial (nē'-mi-al, knē'-mi-al)

cnemidium (nē-mid'-i-um, knē-mid'-i-um)

Cnemidophorus (nē-mi-dof'-ô-rus)

Cnemidotus (nē-mi'-dot-us)

Cneoridum* (nē-ôr-id'-i-um)

Cneorum* (nē-ō'-rum)

Cnethocampa* (nē-thok-am'-pạ)

Cnicus* (nī'-kus)

Cnidaria (nī-dā'-ri-ạ)

Cnidium* (nī'-di-um)

cnidoblast (nī'-dô-blast)

Cnidoscolus* (nī-dô-skō'-lus)

coarctatus (kō-ârk-tā'-tus) pressed together.

coccid (kok'-sid)

Coccidae (kok'-si-dē)

coccigerus (kok-sij'-er-us) berry-bearing.

Coccinellidae (kok-si-nel'-i-dē)

coccineus (kok-sin'-e-us) scarlet.

Coccoloba* (kok-kol'-ô-bạ)

Coccothraustes (kok-ô-thrôs'-tēz)

Cocculus* (kok'-kŭ-lus)

coccus (kok'-us, pl. kok'-sī)

Coccyges (kok-sī'-jēz)

coccyx (kok'-siks)

Coccyzus (kok-sī'-zus)

Cochlearia* (kok-lê-ā'-ri-ạ, kok-lê-âr'-i-ạ)

Codiaeum* (kō-di-ē'-um)

Codonium (kō-dō'-ni-um)

Codonopsis* (kō-dō-nop'-sis)
Coelenterata (sē-len-têr-āt'-ạ)
Coelogenys (sē-loj'-e-nis)
Coelogyne* (sē-loj'-i-nē)
coelom (sē'-lôm, sē'-lom)
coelomic (sē-lō'-mik, sē-lom'-ik)
Coelopleurum* (sē-lô-plŏŏ'-rum)
Coenagriidae (sē-nag-rī'-i-dē)
Coendou (kô-en'-dōō)
coenosarc (sē'-nô-sârk)
coenosus = caenosus (sē-nō'-sus) muddy, foul, dirty.
coerulescens (sē-rul-es'-senz) becoming dark or
    black.
cognatus (kog-nā'-tus) related by blood.
coition (kô-ish'-un)
coitus (kō'-i-tus)
Coix* (kō'-iks)
Colaptes (kô-lap'-tēz)
Colax* (kō'-lax)
Colchicum* (kol'-ki-kum)
Coleogyne* (kol-e-ō'-ji-nē)
Coleonyx (kol-e-on'-iks, kōl-ê-on'-iks)
Coleophora (kol-e-of'-ôr-ạ)
Coleophoridae (kol-e-of-ôr'-id-ē)
Coleoptera (kol-e-op'-ter-ạ, kol-ê-op'-têr-ạ)
coleorhiza (kol-e-ô-rī'-zạ)
Coleus* (kol'-e-ūs, kō'-le-us)
Colinus (kô-līn'-us)
Coliupasser (kō-li- û-pas'-êr)
Colius (kō'-li-us)
collaris (kol-ā'-ris) pertaining to the neck.
collatus (kol-ā'-tus) brought together, gathered.

Collembola (kol-em′-bŏ-lạ)
collenchyma (kol-eng′-ki-mạ)
collencyte (kol′-en-sīt)
Colletes (kol-ē′-tēz)
Colletidae (kol-et′-i-dē)
Collinsia* (kol-in′-si-ạ)
collinus (kol-ī′-nus) pertaining to or of a hill, hilly.
collis (kol′-is) high ground, a hill.
colloid (kol′-oyd)
Collomia* (kol-ō′-mi-ạ)
colludens (kol-ū′-dens) playing together.
Colobus (kol′-ŏ-bus)
Coloptychon (kol-ŏp′-ti-kon)
Coluber (kol′-ŭ-bêr, kol′-u-bêr)
colubrinus (kol-ub-rī′-nus) like a serpent.
colubris (kol-u′-bris) of a serpent.
Columba (kol-um′-bạ)
columbarius (kol-um-bā′-ri-us) of or pertaining to
    a dove.
Columbigallina (kol-um-bi-gal-ī′-nạ)
Colutea* (ko-lūt′-ê-ạ)
Colymbetes (kol-im-bē′-tēz)
Colymbus (kol-im′-bus)
coma (kō′-mạ, ko′-mạ) hair.
Comandra* (kŏ-man′-drạ)
comans (kom′-anz) covered with hair.
Comarum* (kom′-ȧ-rum)
comatosus (kom-at-ō′-sus) hairy.
Comatula (kŏ-mat′-ŭ-lạ)
comes (kō′-mēz)
cometes (kom-ē′-tēz) a comet, also, a proper name.
Commelina* (kom-e-lī′-nạ)

commensal (kom-en'-sal)
commissure (kom'-i-shūr)
communal (kom'-ū-nal)
communis (kom-ū'-nis) general, common.
commutatus (kom-ū-tā'-tus) changed entirely, replaced.
comose (kō'-mōs)
comosus (kom-ō'-sus) furnished with a tuft of hair or leaves, hairy.
compar (kom'-par) equal.
comparative (kom-par'-a-tiv)
complanatus (kom-plā-nā'-tus) level with the ground.
compressus (kom-pres'-us) squeezed together, straight, narrow.
Compsognathus (komp-sog'-na-thus)
Compsothlypidae (komp-soth-lip'-i-dē)
Compsothlypis (komp-soth'-li-pis)
Conandron (kō-nan'-dron)
conarium (kōn-ā'-ri-um)
concha (kong'-ka)
conchiolin (kong-kī'-ŏl-in)
conchology (kong-kol'-ŏj-i)
concinnus (kon-sin'-us) beautiful, striking.
concolor (kon'-ku-lôr) of the same color; also, uniformly colored.
Condalia* (kon-dā'-li-a)
conditor (kon'-di-tôr) a builder, a farmer.
conditus (kon-dī'-tus) seasoned well; also, established.
Condylarthra (kon-di-lâr'-thra)
condyle (kon'-dīl, kon'-dil)

Condylura (kon-dil-ū′-rą)
condylodes (kon-dil-ō′-dēz) knobby, knuckle-like.
Conepatus (kō-ne-pā′-tus)
confertus (kon-fer′-tus) pressed together, crowded, dense.
Confervales* (kon-fêr-vā′-lēz)
confinis (kon-fī′-nis) neighboring, adjoining.
confluens (kon′-flu-enz) running together.
confluentus (kon-flu′-en-tus) croweded together joining, flowing together.
confractus (kon-frakt′-us)
confraternus (kon-frā-têr′-nus) brotherly, also, with affection.
confusus (kon-fū′-sus) confused, perplexed.
conglomeratus (kon-glo-mer-ā′-tus) gathering to form a ball.
congregatus (kon-gre-gā′-tus) collected.
conic (kon′-ik)
conicus (kō′-ni-kus) cone-like.
conidium (kŏn-id′-i-um)
conifer (kō′-ni-fêr, kon′-i-fêr)
Coniferae* (kō-nif′-er-ē)
coniferous (kŏ-nif′-er-us)
Conilurus (kon-i-lū′-rus)
Coniophanes* (kō-ni-ŏ-fā′-nēz)
Coniopteryx (kon-i-op′-tĕ-riks)
Conioselinum* (kō-ni-ŏ-se-lī′-num, kon-i-ŏ-se-lī′-num)
coniospermous (kon-i-ŏ-spêr′-mus)
Conium* (kō-nī′-um)
conjugatus (kon-jug-ā′-tus) united, joined.
connate (kon′-āt)

connexus (kon-eks'-us) joined, cohering.
connivens (kon-ī'-venz) gradually converging.
Connochaetes (kon-ȯ-kē'-tēz)
Connophron (kon'-of-ron, kon-of'-ron)
Conoclinium* (kō-nok-li'-ni-um)
Conolophus (kȯn-ȯl'-of-us)
conopea (kōn-ō'-pe-ạ) resembling a gnat.
Conopholis* (kȯ-nof'-ȯ-lis)
Conopidae (kȯ-nop'-i-dē)
Conopophaga (kȯ-nȯ-pof'-a-gạ)
Conops (kō'-nops)
Conostephium* (kō-nos-tef'-i-um)
consanguineus (kon-sang-win'-e-us) related by blood.
consimilis (kon-si'-mi-lis) similar, like.
consobrinus (kon-sō-brī'-nus) relation, a cousin.
consortes (kon-sôr'-tēz)
contemptus (kon-temp'-tus) despised.
contiguus (kon-ti'-gu-us) near, touching.
continuus (kon-ti'-nu-us) joining, continuous.
Contopus (kon'-tȯ-pus)
contortus (kon-tôr'-tus) twisted.
contractile (kon-trak'-til)
Conuropsis (kon-ûr-op'-sis)
Conurus (kō-nū'-rus)
Conus (kō'-nus)
Convallaria* (kon-val-ā'-ri-ạ)
convallarius (kon-val-ā'-ri-us) forming a valley.
Convolvulus* (kon-vol'-vul-us)
Conyza* (kon-ī'-zạ)
Copaifera* (kȯ-pā-if'-er-ạ)
copepod (kō'-pe-pod)

Copepoda (kō-pep'-ô-dą)
Copidita (kop-id-ī'-tą)
coprolite (kop'-rô-līt)
coprophagus (kop-rof'-å-gus)
Coprotheres (kop-rô-thē'-rēz)
Copsichus (kop'-si-kus)
Coracina (kôr-as-īn'-ą)
Coracius (kôr-ās'-i-us)
Coragyps (kôr'-å-jips)
corallidomous (ko-ral-id'-ô-mus)
corallinus (ko-ral'-in-us) coral-red.
Corallorhiza* (ko-ral-ô-rī'-zą)
corallum (ko-ral'-um)
corbis (kôr'-bis) a basket.
Corchorus* (kôr'-kôr-us)
cordatus (kôr-dā'-tus) wise, prudent.
Corduliidae (kôr-dū-lī'-i-dē)
Cordyline* (kôr-di-lī'-nē)
Cordylophora (kôr-di-lof'-ô-rą)
Coregonus (ko-rē'-gô-nus)
Coreidae (kō-rē'-i-dē)
Corema* (kôr-ē'-mą)
coremata (kôr-ē'-mat-ą)
coremiform (kôr-ē'-mi-fôrm)
coremium (kôr-ē'-mi-um)
Coreopsis* (kôr-ē-op'-sis)
Corethrogyne* (kôr-ē-thro'-ji-nē)
coriaceus (kôr-i-ā'-se-us) made of leather, leathery.
Coriandrum* (kôr-i-an'-drum)
coriifolius (kor-i-i-fol'-i-us, kor-i-i-fō'-li-us) with
　　leathery leaves.
Corisa (kôr'-i-są)

Corispermum* (kôr-i-spêr'-mum)
corium (kō'-ri-um, pl. kō'-ri-ą)
Corixidae (kō-rik'-si-dē)
Corizidae (kôr-iz'-i-dē)
cornea (kôr'-nē-ą)
corniculatus   (kôr-ni-kul-ā'-tus)   horn-shaped, horned.
corniculus (kôr-nik'-ul-us) a small horn.
cornigerus (kôr-nij'-er-us) horn-bearing.
Corningia* (kôr-nin'-ji-ą)
cornubiensis (kôr-nū-bi-en'-sis) of Cornwell.
Cornus* (kôr'-nus)
cornutus (kôr-nū'-tus) horned.
corolla (kȯ-rol'-ą)
coronal (kôr'-ȯ-nal, kȯ-rō'-nal)
coronary (kôr'-ō-nar-i)
coronatus (kôr-ōn-ā'-tus) furnished with a crown.
Coronilla* (kôr-ȯ-nil'-ą)
corporalis (kôr-pôr-āl'-is) pertaining to the body.
corpus (kôr'-pus, pl. kôr'-pôr-ą)
Correa* (kôr'-ê-ą)
Corrigiola* (kôr-ij-i-ō'-lą)
Corrodentia (kôr-ȯ-den'-shi-ą)
Cortaderia* (kôr-tą-dē'-ri-ą)
cortex (kôr'-teks, pl. kôr'-ti-sēz)
Corthylio (kôr-thi'-li-ō)
Corticaria (kôr-ti-kā'-ri-ą)
Corydalidae (kôr-i-dal'-i-dē)
Corydalis (kôr-id'-ȧ-lis)
Corydiidae (kor-i-dī'-i-dē)
Corydon (kor'-i-don)

corylifolius (ko-ri-li-fol'-i-us, ko-ri-li-fō'-li-us) with leaves like the hazel, Corylus.

Corylophodes (kôr-il-of-ō'-dēz)

Corylus* (kôr'-il-us)

corymbose (kôr-im'-bōs)

corymbosus (kor-im-bō'-sus) full of corymbs.

corymbus (kôr-im'-bus)

Corymorpha (kôr-i-môr'-fą)

corynocalyx (kôr-in-ok-āl'-iks) with club-like calyx.

Corynorhinus (kôr-in-ô-rī'-nus)

Corypha* (kôr'-i-fą)

Coryphantha* (kôr-if-an'-thą)

Coryphodon (kôr-if'-ô-don)

Corythaix (ko-rith'-à-iks)

Corythosaurus (kor-ith-ô-sô'-rus)

Corythuca (kôr-ith-ūk'-ą)

Cosmopteryx (koz-mop'-ter-iks)

Cossidae (kos'-i-dē)

Cossus (kos'-us)

Cossypha (kos'-if-ą)

costatus (kos-tā'-tus) ribbed.

Cotinga (kō-ting'-ą)

Cotinus (kot'-in-us)

Cotoneaster* (kô-tō-nê-as'-têr)

Coturnicops (kô-tûr'-nik-ops)

Coturnix (kô-tûr'-niks)

Cotyledon* (kot-i-lē'-dun)

covert (ku'-vêrt)

coxopodite (koks'-ô-pô-dīt)

coxosternum (kok-sôs-têr'-num)

coyote (kō-yō'-te, koy-ō'-tā)

coypu (koy'-po͞o)

Crabronidae (krab-ron'-i-dē)
Cracidae (kras'-i-dē)
Crambe* (kram'-bē)
Crambidae (kram'-bi-dē)
Crambidia (kram-bid'-i-ą)
Crangon (kran'-gon)
craspedum (kras'-pe-dum)
craspedote (kras'-pe-dōt)
crassipes (kras'-i-pēz) fat- or thick-footed.
Crataegus* (krat-ē'-gus)
cratera (krā-tē'-rą)
Crateropus (krat-er'-ô-pus)
craticular (krat-ik'-ul-ar)
Cratogeomys (krat-ô-gē'-ô-mis)
craurus (krô'-rus) brittle.
Creadion (krē-ad'-i-on)
creatine (krē'-ȧ-tin)
crebrus (krē'-brus) close, frequent, repeated.
Creciscus (kres-is'-kus)
cremnobates (krem-nô-bā'-tēz) cliff-climber.
cremocarp (krem'-ô-kârp)
crena (krē'-ną)
crenate (krē'-nāt)
crenatus (kren-ā'-tus) notched.
Crenothrix* (kren'-ô-thriks)
crenulate (kren'-ŭ-lāt)
Creodonta (krē-ô-don'-tą)
Creophilus (krē-of'-il-us)
creper (kre'-per) dusky, dark, doubtful.
Crepidula (krep-id'-ul-ą)
Crepipoda (kre-pip'-od-ą)
Crepis* (krē'-pis)

crepitans (krep'-i-tans) clattering.
crepuscular (krē-pus'-kū-lâr)
Cresentia* (kres-en'-shi-ạ)
Cretaceous (krē-tā'-shus)
Crex (kreks)
cribriform (krib'-ri-fôrm)
Cricetinae (kris-ē-tī'-nē)
Cricetomys (kris-ē'-tŏ-mis)
Cricetus (kris-ē'-tus)
criniflorum (krī-nif-lō'-rum)
crinitus (krī-nī'-tus) covered with hair.
Crinodendron (krīn-od-en'-dron)
Crioceris (krī-os'-e-ris)
Criocerus (krī-os'-er-us)
criocone (krī'-ŏ-kōn)
crispus (kris'-pus) curled, wrinkled, wavy.
Cristatella* (kris-tả-tel'-ạ)
cristatus (kris-tā'-tus) crested, tufted.
crithmoides (krith-mo-ī'-dēz) like *Crithmum.*
Crithmum* (krith'-mum)
Crocanthemum* (krō-kan'-the-mum)
Crocethia (kro-seth'-i-ạ)
croceus (kro'-se-us) of or pertaining to saffron.
Crocidura (kros-id-ū'-rạ)
Crocosmia* (krŏ-koz'-mi-ạ)
Crocus* (krok'-us, krō'-kus)
Crocuta (kro-kū'-tạ)
Cronartium* (krō-nâr'-shi-um)
Crossosoma* (kros-ŏ-sō'-mạ)
Crotalaria* (krot-ả-lā'-ri-ạ)
Crotalus (krot'-ả-lus, krō'-tả-lus)
Crotaphytus (krot-ả-fīt'-us)

Crossosoma. New Latin <Gr. *krossoi*, a fringe+*soma*, a body. Since the penult is long (because it contains a long vowel) it takes the accent. Pronounced: kros-ŏ-sō'-mạ, not krossos'-ŏ-mạ.

Croton* (krōt'-un, krot'-ōn)
Crotophaga (krō-tof'-ả-gạ)
Crucianella* (krū-shi-an-el'-ạ)
Cruciferae* (krū-sif'-êr-ē)
crucis (krū'-sis) of a cross.
cruentus (kru-en'-tus) spotted, stained with blood.
crumena (krū-mē'-nạ)
crus (krūs)
Cryophytum* (krī-o'-fi-tum)
Cryptocercus (krip-tŏ-ser'-kus)
Cryptodira (krip-tŏ-dī'-rạ)
Cryptogramma* (krip-tŏ-gram'-ạ)
Cryptophagus (krip-tof'-ả-gus)
Cryptostegia* (krip-tŏ-stē'-ji-ạ)
Cryptotaenia* (krip-tŏ-tē'-ni-ạ)
Cryptotis (krip-tō'-tis)
Crypturus (krip-tū'-rus)
Cryptus (krip'-tus)
cteinophyte (tīn'-ŏ-fīt)

Ctenium* (ten'-i-um)

ctenocyst (ten-ô'-sist, kten'-ô-sist)

Ctenomys (ten'-ô-mis)

Ctenophora (ten-of'-ô-rạ)

ctenophore (ten'-ô-fôr)

Ctenosaurus (ten-os-ô'-rus)

Ctenucha (ten-ūk'-ạ)

ctetosome (tē'-tô-sōm)

Cucubalus (ku-kū'-bal-us)

Cucujidae (kǔ-kū'-ji-dē)

Cucujus (kū'-kǔ-jus)

cucullatus (kuk-ul-ā'-tus) hooded.

cuculus (ku-kūl'-us) a cuckoo.

Cucumaria (kū-kū-mā'-ri-ạ)

Cucurbita* (kū-kûr'-bi-tạ)

cuirass (kwě-ras')

Culicidae (kū-lis'-i-dē)

cultellus (kul-tel'-us) a little knife.

cultigen (kul'-ti-jen)

Cumingia (kū-min'-ji-ạ)

Cuminum* (kū'-min-um)

cuneatus (kun-e-āt'-us) wedge-shaped.

cuneiform (kǔ-nē'-i-fôrm)

cuneus (ku'-ne-us) a wedge.

cunicularius (kun-ī-kul-ā'-ri-us) a miner.

cuniculus (kun-ī'-ku-lus) a rabbit.

Cunila* (kǔ-nī'-lạ)

Cunonia* (kū-nō'-ni-ạ)

Cuon (kū'-on)

Cupedidae (kū-ped'-id-ē)

Cuphea* (kū'-fě-ạ)

Cupidonia (kū-pi-dō'-ni-ạ)

cupidus (kup'-id-us) a desire, a wish.
cupitus (kup-ī'-tus) desired.
cupreatus (kup-re-ā'-tus) coppery.
Cupressus* (kū-pres'-us)
cuprinus (kup'-rin-us) of copper.
cupule (kū'-pūl)
curassavicus (kū-rá-sav'-i-kus) like the greenish-blue, crested currasow.
Curculionidae (kûr-kū-li-on'-i-dē)
Curcuma* (kûr'-kum-ạ)
Curimatus (kū-ri-mā'-tus)
cursor (kûr'-sôr) a runner.
curtatus (kûr-tā'-tus) shortened.
curtus (kûr'-tus) short, broken, mutilated.
Cuscuta* (kus-kū'-tá, kus'-kŭ-tạ)
cuspidatus (kus-pid-ā'-tus) pointed.
cutaneus (kū-tā'-ne-us) pertaining to or of the skin.
Cuterebra (kū-te-rēb'-rạ, kū-te-reb'-rạ)
cuticle (kū'-tik-l)
cutin (kū'-tin)
Cyamus (sī'-am-us)
Cyanea (sī-ā'-nê-ạ)
cyaneus (sī-an'-e-us) dark-blue, sea-blue.
Cyanocitta (sī-á-nô-sit'-ạ)
Cyanophyceae* (sī-an-ô-fī'-sê-ē)
Cyathea* (si-ath'-ê-ạ)
cyathium (sī-ath'-i-um)
Cyathroceridae (sī-ath-rô-ser'-i-dē)
cyathus (sī'-á-thus)
Cybaeus (sib-ē'-us)
Cybister (sī-bis'-têr)

Cybocephalus (sib-ȯ-sef'-al-us)
cycad* (sī'-kad)
Cyclamen* (sī'-kla-men, sik'-lȧ-men)
Cyclanthera* (sī-klan-thē'-rạ, sik-lan'-thê-rạ)
Cyclaris (sik'-lä-ris)
cyclic (sik'-lik, sī'-klik)
cyclocerculus (sīk-klȯ-ser'-ku-lus)   round+tail+
  -ulus, a diminutive ending.
cycloid (sī'-kloid)
Cycloloma* (sī-klō-lōm'-ạ, si-klō-lō'-mạ)
Cycloplasis (sī-klop-lās'-is)
Cyclophorus (sī-klof'-ȯ-rus)
Cyclorhapha (sī-klȯr'-ȧ-fạ)
cyclosis (sī-klō'-sis) a whirling, a circulation.
Cyclostomata (sī-klȯ-stom'-at-ạ)
cyclostomate (sī-klos'-tȯ-māt)
Cyclostrema (sī-klȯ-strē'-mạ)
Cyclothurus (sī-klȯ-thū'-rus)
Cyclotus (sī-klō'-tus)
Cyclura (sī-kloo͞'-rạ)
Cyclas (sī'-klas)
cydariform (sī-dar'-i-fôrm)
Cydippe (sī-dip'-ē)
Cydippida (sī-dip'-i-dạ)
Cydnidae (sid'-ni-dē)
Cydonia* (sī-dō'-ni-ạ)
cyesis (sī-ē'-sis)
Cygnopsis (sig-nop'-sis)
Cygnus (sig'-nus)
Cylas (sī'-las)
Cylichna (sil-ik'-nạ)
Cylindroleberis (sil-in-drȯ-leb'-er-is)

Cyllene (si-lē'-nē)
Cymatogaster (sī-mat-ȯ-gas'-têr)
Cymatophora (sī-mat-of'-ôr-ạ)
Cymbidium* (sim-bid'-i-um)
Cymbopogon* (sim-bȯ-pō'-gōn)
Cymindis (sim-in'-dis)
Cymopterus* (sī-mop'-ter-us)
cymose (sī'-mōs, sī-mōs')
Cymothoa (sī-moth'-ȯ-ạ)
Cymothoidae (sī-mo-thō'-id-ē)
Cynaelurus (sī-nē-lū'-rus)
Cynanchum* (sin-ang'-kum)
Cynanthus* (sin-an'-thus)
Cynara* (sin'-a-rạ, sin'-âr-ạ)
Cynictis (sī-nik'-tis)
Cynipidae* (sin-ip'-i-dē)
Cynips* (sin'-ips, sī'-nips)
Cynoctonum* (sin-ok'-ton-um)
Cynodon* (sin'-ȯ-don, sī'-nod-on)
Cynodonta (sin-ȯ-dont'-ạ, sī-nȯ-dont'-ạ)
Cynogale (sin-og'-al-ē)
Cynoglossum* (sin-ȯ-glōs'-um, sī-nȯ-glos'-um)
Cynomys (sin'-ȯ-mis)
Cynopithecus (sin-ȯ-pi-thē'-kus, sī-nȯ-pi-thē'kus)
Cynosurus* (sin-ȯ-sū'-rus, sī-nȯ-sū'-rus)
Cyperus* (sip-ē'-rus, sī-pē'-rus)
cyphella (sif-el'-ạ, sī-fel'-ạ)
Cyphomandra* (sī-fom-an'-drạ)
Cyphon (sī'-fon)
Cyphornis (sif-ôr'-nis)
Cypraea (sip-rē'-ạ, sī-prē'-ạ)
Cyprinus (sip-rī'-nus)

Cypripedium* (sip-rip-ed'-i-um, sip-ri-pē'-di-um)
cypsela (sip'-sel-ạ)
Cypselus (sip'-se-lus)
Cyrilla* (sī-ril'-ạ)
Cyrtonotum (sêr-ton-ōt'-um)

Cynomys. New Latin <Gr. *kyōn*, *kynos*, dog + *mys*, a mouse. The generic name of the prairie dog or "the rodent that barks like a dog." Accent on the first syllable. Pronounced: sin'-ō-mis, not sin-ō'-miz.

Cyrtonyx (sêr-tō'-niks)
Cyrtophium (sêr-tof'-i-um)
Cyrtopogon (sêr-tồ-pō'-gōn)
Cystacanthus* (sis-tak-anth'-us)
Cystignathus (sis-tig'-na-thus)
Cystophora (sis-tof'-ôr-ạ)
Cystopteris* (sis-top'-têr-is)
Cystopus* (sis'-top-us)
Cytherea (sith-e-rē'-ạ)
Cytinus* (sit'-i-nus)
Cytisus* (sit'-i-sus)
Cytophyllum* (sit-of-il'-um)
cytula (sit'-ūl-ạ)

D

Daboecia* (dȧ-bō-ē′-shi-ạ)
Dacelo (dȧ-sē′-lō)
Dacnusa (dak-nū′-sạ)
dacryocyst (dak′-ri-ȯ-sist)
Dactilomys (dak-til′-ȯ-mis)
dactyl (dak′-til)
Dactylis* (dak′-til-is)
dactyloides (dak-til-o-ī′-dēz) finger-like.
Dactylopius (dak-til-ōp′-i-us)
Daemonelix (dē-mon′-e-liks)
daemonius (dē-mon′-i-us) elfin, strange, marvelous.
Dafila (daf′-i-lạ)
Dahlia* (dä′-li-ạ)
Dalbergia* (dal-bêr′-gi-ạ)
Dalcerides (dal-ser′-id-ēz)
Dalibarda* (dal-i-bâr′-dạ)
dama (dā′-mạ) a fallow deer.
Damalis (dam′-ȧ-lis)
Damaliscus (dam-al-is′-kus)
damnosus (dam-nō′-sus) injurious, destructive.
Danaidae (dā-nā′-i-dē)
Danaus (dā′-nȧ-us)
Danthonia* (dan-thō′-ni-ạ)
Daphne* (daf′-nẽ)
daphnoides (daf-no-ī′-dēs) laurel-like.
Daptrius (dap′-tri-us)
darnel (dâr′-nel)
dartos (dâr′-tos)
Dascyllus (da-sil′-us)
Dasyatidae (das-i-at′-i-dē)

dasyclados (das-ik'-la-dos) shaggy-twigged.
Dasylirion (das-i-lī'-ri-on, das-i-lir'-i-on)
Dasymys (das'-i-mis)
dasypaedes (das-i-pē'-dēz)
Dasypeltis (das-i-pel'-tis)
dasyphyllus (das-if-il'-us) shaggy or hairy leafed.
Dasyprocta (das-i-prok'-tạ)
Dasypus (das'-i-pus)
Dasyurus (das-i-ūr'-rus)
Datana (dā-tā'-nạ)
datum (dā'-tum, pl. dā'-tạ)
Datura* (da-tū'-rạ)
Daucus* (dô'-kus)
daunus (dô'-nus) fabled king of part of Apulia.
Davallia* (dav-al'-i-ạ)
dealbatus (de-al-bā'-tus) whitened, plastered with white-wash.
debilis (dē'-bi-lis) crippled, feeble, weak.
decapetalus (dek-ap-et'-al-us) ten-petaled.
Decapoda (de-kap'-ŏ-dạ)
Decatoma (de-kat'-ŏm-ạ)
decemjugate (des-em-jū'-gāt)
decens (de'-senz) seemly, fit, well-formed.
deceptus (dē-sep'-tus) deceiving.
decidua (dē-sid'-ŭ-ạ)
decipiens (dē-sip'-i-enz) deceiving.
decisum (dē-sī'-sum) settled, determined.
declinatus (dē-klī-nā'-tus) bent aside, turned down.
declivis (dē-klī'-vis) sloping.
Decodon* (dek'-ŏ-don)
decollatus (dē-kol-ā'-tus) beheaded.
decolorans (dē-kol'-ôr-anz) without color.

decorus (dek-ōr'-us) elegant, becoming.
Decumaria* (dek-ū-mā'-ri-ą)
decumbens (dē-kum'-benz) lying down.
decurrens (dē-kêr'-enz) •
decussate (dek'-us-āt, dē-kus'-āt)
decussatus (dek-us-ā'-tus) divided crosswise.
defecate (def'-e-kāt)
dehiscence (dē-his'-ens)
dehiscent (dē-his'-ent)
deirids (dī'-ridz)
Deirochelys (dī-rok'-e-lis)
deletrix (dē-lē'-triks) she that destroys.
deletus (dē-lē'-tus) abolished, finished.
delicatus (dē-lik-ā'-tus) alluring, delightful.
Delonix* (dē-lō'-niks)
Delostoma* (dē-los'-tom-ą)
Delphacidae (del-fas'-i-dē)
Delphinapterus (del-fin-ap'-têr-us)
Delphinium* (del-fin'-i-um)
Delphinus (del-fī'-nus)
deltoides (del-to-ī'-dēz)
deltoideus (del-toyd'-e-us) delta-like.
deme (dēm)
demersed (dē-mêrst')
demersus (dē-mêr'-us) submerged.
demissus (dē-mis'-us) low-lying, hanging down.
demorsus (dē-môr'-sus) bitten off.
Dendragapus (den-drag'- à-pus)
Dendraspis (den-dras'-pis)
Dendrobates (den-drob'-à-tēz)
Dendrobium* (den-dro'-bi-um)
Dendrocalamus* (den-drô-kal'-am-us)

Dendroctonus (den-drok'-tôn-us)
Dendrohyrax (den-drô'-hi-raks)
Dendroica (den-droy'-kạ)
Dendroides (den-dro-īd'-ēz)
Dendrolagus (den-drō'-lag-us)
Dendroleon (den-drô-lē'-ōn)
Dendromecon* (den-drom-ē'-kon)
Dendromys (den'-drô-mis)
dendron (den'-dron)
Dendropanax* (den-drop'-an-aks)
Dendrophidion (den-drô-fid'-i-on)
Dendroseris* (den-dros'-er-is)
Dennstaedtia* (den-stēdt'-i-ạ)
densleonis (denz-lē-ōn'-is) lion's tooth.
densus (den'-sus) thick, dense, set close.
Dentalium (den-tā'-li-um)
Dentaria* (den-tā'-ri-ạ)
dentatus (den-tā'-tus) having teeth.
denticulatus (den-tik-ul-ā'-tus) having small teeth.
denudate (v. den'-û-dāt, dê-nūd'-āt; adj. dê-nūd'-
    āt, den'-û-dāt)
denudatus (dē-nu-dā'-tus) stripped, laid bare.
depictus (dē-pik'-tus) portrayed, described.
depilans (dē'-pi-lanz) despoiling of feathers or
    hair, making bald.
depauperatus (dē-pô-pêr-ā'-tus) stunted, having a
    poor appearance.
dentaneus (den-tā'-ne-us) threatening.
depilation (dep-i-lā'-shun)
depula (dep'-ûl-ạ)
derelictus (dē-re-likt'-us) abandoned, neglected.
dermalia (dêr-mā'-li-ạ)

Dermaptera (dêr-map′-têr-ạ)
Dermestidae (dêr-mes′-ti-dē)
Dermochelys (dêr-mok′-e-lis)
Derotremata (der-ŏ-trē′-mat-ạ)
dertrotheca (der-trŏ-thē′-kạ)
deserti (dez-êr′-tī) of the desert.
deserticolous (dez-êr-tik′-o-lus)
desiccant (des′-i-kant, dê-sik′-ant)
Desmana (des′-man-ạ)
Desmanthus* (des-man′-thus)
Desmodium* (des-mō′-di-um)
Desmodus (des′-mŏ-dus)
Desmognathus (des-mog′-na-thus)
Desmoncus* (des-mon′-kus)
desquamate (des′-kwa-māt, dê-skwā′-māt)
detonsus (dē-ton′-sus) clipped off, sheared.
detritus (dē-trī′-tus) worn out, trite.
deustus (de-us′-tus) consumed, burned up, destroyed.
deuteroplasm (dū′-têr-ŏ-plazm)
deutoplasm (dū′-tŏ-plazm)
Deutzia* (dū′-tzi-ạ)
devexus (dē-veks′-us) inclining downwards, steep.
Devonian (de-vō′-ni-an)
diabetes (dī-ȧ-bēt′-ēz)
diabetic (dī-ȧ-bet′-ik)
diabolis (di-ab′-ol-is)
Diacalpe (di-ak-al′-pē)
Diacrisia (dī-ak-ris′-i-ạ)
Diacrium (dī-ak′-ri-um)
Diadasia (dī-ad-ās′-i-ạ)
Diadophis (dī-ad′-ŏf-is)

Diamorus (dī-am′-ôr-us)
Dianthera* (dī-an′-thêr-ạ, di-an-thē′-rạ)
Dianthus* (dī-an′-thus, di-an′-thus)
diapedesis (dī-ạ-ped-ēs′-is)
Diapensia* (dī-ạ-pen′-si-ạ)
Diaperis (dī-ap-ēr′-is)
Diaphania (dī-af-ān′-i-ạ)
Diaphora* (dī-af′-ōr-ạ)
diaphysis (dī-af′-i-sis)
Diapria (dī-ap′-ri-ạ)
diarch (dī′-ârk)
Diarrhena* (dī-ä-rhēn′-ạ)
Diascia* (dī-as′-ki-ạ)
Diaspididae (dī-as-pid′-i-dē)
Diaspinae (dī-as-pī′-nē)
Diastata (dī-as′-tạ-tạ)
diathetic (dī-ạ-thet′-ik)
diatom (dī′-ạ-tom, di′-ạ-tōm)
Diatraea (dī-at-rē′-ạ)
Diatropura (dī-at-rop′-ûr-ạ)
Diatryma (dī-at-rī′-mạ)
Dibelodon (dī-bel′-ô-don)
Dicaeum (dī-sē′-um)
Dicamptodon (dī-kamp′-tô-don)
Dicentra* (dī-sen′-trạ, dis-en′-trạ)
Diceras (dis′-e-ras)
Diceratherium (dis-êr-ạ-thē′-ri-um)
Diceros (dis′-êr-os)
dichasium (dī-kā′-zi-um)
Dichelesthium (dī-kel-es′-thi-um)
Dichelostemma* (dī-kel-os-tem′-ạ)
Dichelyma (dī-kel′-i-mạ)

dichogamy (dī-kog'-am-i)
Dichondra* (dī-kon'-drạ)
Dichromanassa (dī-krō-man-as'-ạ)
diclinic (dī'-klin-ik)
diclinous (dī'-klin-us)
Dicoria* (dī-kôr'-i-ạ)
Dicotyles (dī-kot'-i-lēz)
Dicrostonyx (dī-kros'-tŏ-niks)
Dictamnus* (dik-tam'-nus)
didactic (di-dak'-tic, dī-dak'-tic)
Didelphys (dī-del'-fis)
Didineis (dī-din'-e-is)
Didinium (dī-din'-i-um)
Didiplis* (di-di'-plis)
Didunculus (did-ung'-kŭ-lus)
Didymocarpus* (did-im-ok-âr'-pus)
didymus (did'-i-mus) in pairs.
didynamous (dī-din'-a-mus)
Diedrocephala (dī-ed-rŏ-sef'-al-ạ)
Dielasma (dī-ê-las'-mạ)
Diemictylus (dī-em-ik'-ti-lus)
Dierama* (di-er-ā'-mạ)
Diervilla* (dī-êr-vil'-ạ)
Difflugia (dif-lū'-ji-ạ)
Digenea (dī-jen'-ê-ạ, dī-jēn'-ê-ạ)
digenous (dij'-en-us)
Digera* (dij'-er-ạ)
digestion (dī-jest'-chun)
Digitalis* (dij-i-tā'-lis)
Digitaria* (dij-i-tā'-ri-ạ)
digitatus (dij-it-ā'-tus) with fingers.
digitigrade (dij'-it-i-grād)

Diglochis (dī-glō′-kis)
dignabilis (dig-nā′-bil-is) worthy.
Digraphis* (dig′-raf-is)
digynous (dij′-in-us, dī′-jin-us)
dihybrid (dī-hī′-brid)
Dilaridae (dī-lar′-i-dē)
dilatation (dil′-a̦-tā′-shun, dī-la̦-tā′-shun)
dilatatus (dī-lā-tā′-tus) spread out, enlarged.
dilate (dī-lāt′, di-lāt′)
dilectus (dī-lek′-tus) precious, valuable.
Dilophus (dil′-of-us, dī′-lof-us)
dilute (di-lūt′, dī-lut′)
dilutior (dī-lū′-ti-ôr) thinner, weaker, softer.
Dimecodon (dī-mē′-kô-don)
dimeric (dī-mer′-ik)
dimerous (dim-êr′-us)
dimerus (dim′-er-us) in two parts.
dimidiate (dī-mid′-i-āt, dim-id′-i-āt)
dimidiatus (dī-mi-di-ā′-tus) halved, half.
Dimorphotheca* (dī-môrf-ô-thē′-ka̦)
Dinacrida (dī-nak′-ri-da̦)
Dinetus (dī-nē′-tus)
Dineutes (din-ū′-tēz)
Dinoceras (dī-nos′-er-as)
Dinoflagellata (din-ô-flaj-e-lāt′-a̦)
dinomic (dī-nom′-ik)
Dinomys (dī′-nô-mis)
Dinophilea (dī-nof-i-lē′-a̦)
Dinopidae (dī-nop′-i-dē)
Dinopis (dī-nō′-pis)
dinosaur (dī′-nō-sôr)
Dinotherium (dī-nō-thē′-ri-um)

Dinohyus (dī-nō-hī'-us)
Diodia* (dī-ō'-di-ạ)
dioecius (dī-ē'-shus, dĭ-ē'-si-us)
dioestrus (dī-ē'-strus)
dioicous (dī-oy'-kus)
Diomedea (di-o-mē-dē'-ạ)
diomedeus (di-o-mēd'-e-us) Diomedes, hero at the
siege of Troy.
Dionaea* (dī-ō-nē'-ạ, di-ō-nē'-ạ)
Dioön (dī-ō'-on)
Diopogon* (dī-ọ-pōg'-ōn)
Diornis (dī-ôrn'-is)
Dioscorea* (dī-os-kọ-rē'-ạ, di-os-kor'-e-ạ)
Diosma* (dī-oz'-ma, dī-os'-mạ, di-os'-mạ)
Diospyros* (dī-os'-pi-ros)
Diotis* (di-ō'-tis)
Diphyes (dif'-i-ēz)
Diphylleia* (dī-fi-lē'-yạ, dif-i-lē'-yạ)
Diphyllobothrium (dī-fil-ọ-both'-ri-um, dif-il-ọ-
both'-ri-um)
diphyllus (dif-il'-us) two-leaved.
diphyodont (dif'-i-ọ-dont)
Diplodocus (dip-lod'-ọ-kus)
diploë (dip'-lọ-ē)
Diploglossata (dip-lọ-glos'-at-ạ)
diploid (dip'-loyd)
Diplomys (dip'-lọ-mis)
Diplophysa (dip-lọ-fī'-sạ)
Diploplectron (dip-lọ-plek'-tron)
Diplopoda (dip-lop'-ọ-dạ)
dipnoan (dip'-nọ-an)
Dipnoi (dip'-nọ-ī)

Diplodocus <Gr. *diplo*- <*diploos*, double, twofold+*dokos*, a main beam or bar. Pronounced: di-plod'-ŏ-kus, not dip-lō-dō'-kus.

Dipodomys (dī-pod'-ŏ-mis)
Diprion (dip-rī'-on)
Diprionidae (dip-ri-ō'-ni-dē)
diprotodont (dī-prō'-tŏ-dont)
Dipsacus* (dip'-sa-kus)
Dipsas (dip'-sas)
Dipsosaurus (dip-sō-sô'-rus)
Diptera (dip'-têr-ạ)
Dipylidium (dī-pil-id'-i-um)
Dirca* (dêr'-kạ)
Dircaea (dêr-sē'-ạ)
Dircenna (dêr-sen'-ạ)
dirus (dī'-rus) dreadful, awful, ill-omened.
Disarenum* (dis-ar'-en-um)
Dischidia* (dis-kid'-i-ạ)
discors (dis'-kôrs) disagreeing.
disideratus (dis-īd-er-ā'-tus) twice sunstruck.
disjunctus (dis-junkt'-us) separated, distant, remote, disjoined.
dispar (dis'-pâr) unlike, different, unequal.

dispermic (dī-spêr′-mik)
Dispholidus (dis-fol′-id-us)
Disporum* (dī-spō′-rum, dis′-pôr-um)
dissect (di-sekt′)
dissectus (dis-ekt′-us) deeply cut.
dissitus (dis′-it-us) remote.
Dissosteira (dis-os-tī′-ra)
Dissoura (dis′-ûr-a)
distachyon (dis-tak′-i-on) two-spiked.
distachyus (dis-ta′-ki-us) two-spiked.
distans (dis′-tans) standing apart, distant.
Distichlis* (dis-tik′-lis)
distichus (dis′-tik-us)
districhum (dis′-trik-um)
distylus (dis′-til-us)
Dithyrea* (dith-i-rē′-a)
Ditoma (dit′-ȯ-ma)
Ditrocha (dit′-rȯ-ka)
ditrochous (dit′-rȯ-kus)
Diuris* (di-ū′-ris)
diurnal (dī-ûr′-nal)
diurnus (di-ûr′-nus) belonging to the day, of the
     day.
divaricate (dī-var′-i-kāt)
divaricatus (dī-vār-i-kā′-tus) spread apart.
divergens (dī-ver′-jenz) bending, inclining away
     from.
divergent (dī-ver′-jent)
dives (dī′-vēz) rich, splendid, precious.
divisus (dī-vī′-sus) divided.
divulsus (dī-vul′-sus) rent asunder, torn, sepa-
     rated.

Dizygotheca* (dī-zī-gồ-thē'-kạ)
dodecagynous (dō-de-kaj'-i-nus)
Dodecatheon* (dō-de-kath'-e-on)
Dodonea* (dō-dồ-nē'-ạ, dod-on-ē'-ạ)
Dohrniphora (dôr-nif'-ôr-ạ)
dolabratus (dol-ā-brā'-tus) shaped like a pick-ax.
Dolichoglossus (dol-ik-ồ-glos'-us)
Dolicholus* (dol-ik'-ol-us)
Dolichonyx (dol-ik'-ồ-niks)
Dolichopsyllidae (dol-i-kồ-psil'-i-dē)
Dolichos* (dol'-i-kos)
Dolichotis (dol-i-kō'-tis)
Doliolum (dō-lī'-ồ-lum)
Dolium (dō'-li-um)
Dombeya* (dom-be'-ạ)
domesticus (dom-es'-ti-kus) belonging to one's
    family or household.
domicile (dom'-i-sil)
Dominula (dom-in'-ul-ạ)
donax (dō-naks', don'-aks) a sort of reed; also,
    the male scallop or pecten.
Dondia* (don'-di-ạ)
Doris (dō'-ris)
Doronicum* (dō-rō-nī'-kum, dō-ron'-ik-um)
Dorosoma (dôr-ồ-sō'-mạ)
dorsalis (dôr-sā'-lis) pertaining to or of the back.
Doryanthes* (dôr-i-anth'ēz)
Dorycnium* (dôr-ik'-ni-um)
Dosinia (dō-sin'-i-ạ)
dovekie (duv'-ki)
Dovyalis* (dō-vi'-āl-is)
Dowingia* (dow-inj'-i-ạ)

dowitcher (dow'-ich-êr)
Downingia* (down-inj'-i-ạ)
Doxantha* (doks-an'-thạ)
Draba* (drā'-bạ)
Dracaena* (drȧ-sē'-nạ)
Dracocephalum* (drak-ỏ-sef'-al-um)
Draconis* (drak-ō'-nis)
Dracontium (drak-on'-ti-um, drak-on'-shi-um)
Drapetes (drȧ-pēt'-ēz)
Drassus (dra'-sus)
Drasterius (dras-tē'-ri-us)
Dreissena (drī'-se-nạ)
Drepana (drep'-ȧ-nạ)
Drepane (drep'-ȧ-nē)
Drepanis (drep'-ȧ-nis)
drepanophyllus (drep-an-of-il'-us) with sickle-
shaped leaves.
Drimys* (drī'-mis)
Driosporos* (drī-os-pō'-ros)
Dromaeus (drỏ-mē'-us)
Dromas (drō'-mas)
dromedary (drom'-ė-der-i)
Dromiacea (drỏ-mi-ā'-sė-ạ)
Dromicia (drō-mish'-i-ạ)
Dromocyon (drō-mō'-si-on)
Drosera* (dros'-er-ạ)
Drosophila (drỏ-sof'-il-ạ)
Drosophilidae (dros-ỏ-fil'-i-dē)
drupe (drūp)
Dryadophis (drī-ad-ōf'-is)
Dryas* (drī'-as)
Dryinidae (drī-in'-i-dē)

Dryinus (drī'-in-us)
Drymarchon (drī-mâr'-kōn)
Drymnobius (drim-nō'-bi-us)
Drymomys (drim'-ȯ-mis)
Drynaria* (drī-nā'-ri-ạ)
Dryobates (drī-ȯ-bā'-tēz)
Dryocopus (drī-ok'-ȯ-pus)
Dryopetalon* (drī-ȯ-pet'-al-on)
Dryopithecus (drī-ȯ-pi-thē'-kus)
Dryopteris* (drī-op'-ter-is)
dubius (dub'-i-us) fluctuating, undecided, moving
  in two ways.
dugong (dū'-gong)
duiker (dī'-kêr)
dulcamara (dul-ka-mä'-rạ)
dulosis (dū-lō'-sis)
Dulus (dū'-lus)
Dumetella (dū-mēt-el'-ạ)
dumetorum (dū-mē-tō'-rum) of thickets.
duodenal (dū-ȯ-dē'-nal)
duodenum (dū-od-ē'-num, dū-ȯ-dē'-num)
duramen (dū-rā'-men)
Durio* (dū'-ri-ō)
duriusculus (dū-ri-us'-ku-lus) somewhat hard.
Dyctina (dict'-in-ạ)
Dysdera (dis'-der-ạ)
Dysdercus (dis-der'-kus)
Dysodia* (dis-ō'-di-ạ)
Dyssochroma (dis-sok-rō'-mạ)
Dyssodia* (dis-sō'-di-ạ)
dystrophic (dis-trof'-ik)
Dytiscidae (dī-tis'-i-dē)

E

Eacles (ē'-à-klēz)
Earina* (ē-ar'-in-ạ)
ebracteatus (ē-brak-te-ā'-tus) without bracts.
ebrius (ēb'-ri-us) drunken.
Eburia (ē-bū'-ri-ạ)
eburneus (e-bûr'-ne-us) of ivory.
ecalcaratus (ē-kal-kar-ā'-tus) without spurs.
Ecballium* (ek-bal'-i-um)
Eccremocarpus* (ek-rem-ȯ-kâr'-pus)
ecderon (ek'-dêr-on)
ecdysis (ek'-dī-sis)
Ecdyuridae (ek-dī-ūr'-i-dē)
echard (ek-ârd')
Echeveria* (ek-ev-ē'-ri-ạ)
Echimys (ek-i'-mis)
Echinacea* (ek-ī-nā'-se-ạ)
echinatus (ek-īn-ā'-tus) prickly, spiny.
Echinochloa* (ek-ī-nok'-lȯ-ạ)
Echinococcus (ek-ī-nȯ-ko'-kus)
Echinocystis* (ek-ī-nȯ-sis'-tis)
Echinodermata (ek-īn-ȯ-dêr'-màt-ạ)
Echinodorus* (ek-ī-nȯ-dō'-rus)
Echinoidea (ek-in-oy'-dē-ạ)
echinoides (ek-ī-no-ī'-dez) hedgehog-like, prickly.
Echinophora* (ek-īn-of'-ō-rạ)
Echinophthiriidae (ek-ī-nof-thir-ī'-i-dē)
Echinops* (ek-ī'-nops)
Echinopsis* (ek-ī-nop'-sis)
Echinosorex (ek-ī-nȯ-sō'-reks)
Echinospermum* (ek-ī-nos-pêr'-mum)
echinulate (ek-in'-û-lāt)

Echioglossum* (ek-i-ô-glos'-um, ek-i-ô-glōs'-um)
Echis (ek'-is)
Echites* (ek-ī'-tēz)
Echium* (ek'-i-um)
Echiurus (ek-i-ū'-rus)
Eciton (es'-i-ton)
Eclipta* (ē-klip'-tạ)
eclosion (êk-lō'-zhun)
ecobiotic (ē-kô-bī-ot'-ik)
ecology (ē-kol'-ô-ji)
Ectobia (ek-tō'-bi-ạ)
Ectrichodia (ek-tri-kō'-di-ạ)
edaphic (e-daf'-ik)
edaphon (ed'-af-on)
edax (e'-daks) gluttonous.
edentulus (ē-dent'-u-lus) toothless.
Edraianthus* (ed-rā-i-an'-thus)
Edriaster (ed-ri-as'-têr)
edulis (ed-ū'-lis) edible.
effector (ef-ekt'-ôr, ef-ekt'-êr)
efferent (ef'-er-ent)
efferus (ef'-er-us) wild, fierce.
efficax (ef'-i-kaks) powerful, efficient.
effusus (ef-ū'-sus) loose-spreading.
Ega (ēg'-ạ)
egenus (ej-ē'-nus) needy, in want of, poor, worth-
less.
egg (eg)
Eglanteria* (eg-lan-tē'-ri-ạ)
egret (ē'-gret, eg'-ret)
Egretta (ē-gret'-ạ)
Eichhornia* (īk-hôr'-ni-ạ)

Elachista (el-a-kis'-tạ)
Elachistodon (el-a-kist'-ô-don)
Elaeagnus* (el-ê-ag'-nus)
Elaeis* (ê-lē'-is)
elaeocyte (el-ē'-ô-sīt)
elaioplast (el-ī'-ô-plast)
Elanoides (el-a-no-ī'-dēs, el-a-noy'-dēs)
Elanus (el'-ạ-nus)
Elaphe (el'-a-fē)
Elaphodus (e-laf'-ô-dus)
Elaphoglossum* (el-af-og-lōs'-um, el-af-og-los'-um)
Elaphrium* el-af'-ri-um)
Elaphrus (el-af'-rus)
Elaphus (el'-a-fus)
Elaps (ē'-laps)
Elasmognathus (el-as-mog'-na-thus)
elassodon (el-as'-ô-don) a driving tooth.
Elassoma (el-ạ-sō'-mạ)
elater (el'-ạ-têr)
Elateridae (el-ạ-ter'-i-dē)
Elatine* (el-at-ī'-nē)
elatior (e-lā'-ti-or)
elatus (ē-lā'-tus) high, tall.
Eledone (el-e-dō'-nē)
Eleocharis (el-ē-ok'-ạ-ris)
Eleodes (el-ê-ō'-dēz)
Eleotris (el-ê-ō'-tris)
Elephantopus* (el-e-fan'-tô-pus)
Elephas (el'-e-fas)
Eleusine* (el-ū-sī'-nē)
Eleutherurus (el-ū-the-rū'-rus)
Elgaria (el-gā'-ri-ạ)

eligulate (ē-lig'-ū-lāt)
Elis (ē'-lis)
Elodea* (ê-lo'-dê-ạ, el-ô-dē'-ạ)
elongatus (ē-lon-gā'-tus) removed, kept aloof.
Elops (el'-ops)
Elymus* (el'-i-mus)
Elysia (ê-lis'-i-ạ)
Elytraria* (el-ī-trā'-ri-ạ)
elytrum (el'-i-trum)
Emballonura (em-bal-ô-nū'-rạ)
Emberiza (em-ber-ī'-zạ)
Embernagra (em-bêr-nā'-grạ)
Embiidae (em-bī'-i-dē)
Embioptera (em-bi-op'-te-rạ)
embryo (em'-bri-ō)
embryonal (em-bri-ōn'-al)
emendation (ē-men-dā'-shun, em-en-dā'-shun)
Emerita (ê-mer'-i-tạ)
Emesa (em'-es-ạ)
Emesis (em'-e-sis)
eminens (ēm'-i-nenz) prominent, lofty.
Empetrum* (em-pet'-rum)
Emphytus* (em'-fit-us)
Empididae (em-pid'-i-dē)
Empidonax (em-pid'-ô-naks)
Emyda (em'-i-dạ)
Emys (e'-mis)
Enaliornis (en-al-i-ôr'-nis)
Enallagma (en-al-ag'-mạ)
enantius (en-an'-ti-us) opposite.
encaustus (en-kô'-stus) burned in.
Encelia* (en-sēl'-i-ạ, en-sel'-i-ạ)

Enceliopsis* (en-sēl-i-op′-sis)
Enchelys (en′-ke-lis)
Enchenopa (en-ken-ō′-pạ)
enchylema (eng-kīl-ē′-mạ)
Enchytraeus (eng-ki-trē′-us)
Encope (en′-kŏ-pē)
Encrinus (en′-kri-nus)
Encyrtidae (en-sêr′-ti-dē)
endemic (en-dēm′-ik, en-dem′-ik)
endocrine (en′-dŏ-krīn, en′-dŏ-krin)
endogenous (en-doj′-e-nus)
endognathal (en-dog′-nạ-thal)
endolysin (en-dol′-is-in)
Endomychidae (en-dŏ-mik′-idē)
Endomychus (en-dom′-i-kus)
endopodite (en′-dop-ŏ-dīt, en-dop′-ŏ-dīt)
Endymion* (en-dim′-i-on)
energid (en-êr′-jid)
enerterus (en-er′-ter-us)
Engystomatidae (en-ji-stōm-at′-i-dē)

Enhydra. The generic name of the sea-otter. <Gr. *enydris*, the otter <*enydros*, living in water. The accent is on the antepenult. Pronounced: en′-hi-drạ.

Enhydra (en'-hi-drą)
Enicocephalidae (en-i-kô-se-fal'-i-dē)
enixus (ē-niks'-us) ascending, bringing forth.
enneaphyllus (en-ê-a-fil'-us) nine-leaved.
Ennearthron (en-ê-âr'-thron)
Enneopogon* (en-ê-o-pōg'-ōn)
Enoclerus (en-ok-lē'-rus)
Enodia (en-ōd'-i-ą)
Enophrys (e-nof'-ris)
Ensatina (en-sāt-īn'-ą)
ensatus (en-sāt'-us) sword-like.
Ensete* (en-sē'-tē)
ensifolia (en-si-fol'-i-ą, en-si-fō'-li-ą) with sword-like leaves.
Entemobryidae (en-tem-ôb-rī'-i-dē)
enteron (en'-ter-on)
Entomostraca (en-tô-mos'-trȧ-ką)
Entosphenus (en-tô-sfē'-nus)
enucleator (e-nū-kle-ā'-tôr) a taker out of kernels, one who shells nuts.
enzyme (en'-zīm, en'-zim)
Eoanthropus (ē-ô-an-thrō'-pus)
Eocene (ē'-ô-sen)
Eogaea (ē-ô-jē'-ą)
Eohippus (ē-ô-hip'-us)
Eois (ē-ō'-is)
Eopsaltria (ē-ō-sol'-tri-ą)
eos (ē'-os) sunrise.
Eosentomidae (ē-ōs-en-tom'-i-dē)
Eosentomon (ē-ō-sen'-to-mon)
Epacris* (ep'-ȧ-kris, ep-ak'-ris)
Epeira (ep-ī'-rą)

Ephedra <L. *ephedra*, the horsetail
<Gr. *ephedra* <*ephedros*, sitting upon.
Pronounced: ef-e'-dra̧; the Century
Dictionary places the accent upon the
first syllable, ef'-e-dra̧.

Epeiridae (ĕ-pī'-ri-dē)
Ephedra* (ef-ed'-ra̧)
Ephemerellidae (ef-e-mer-el'-i-dē)
ephemerid (ef-em'-e-rid)
Ephemeridae (ef-e-mer'-i-dē)
Ephestia (ef-es'-ti-a̧)
Ephydra (ef'-īd-ra̧)
ephydrid (ef'-i-drid)
ephyra (ef'-i-ra̧)
Epicauta (ep-i-kô'-ta̧)
epichilium (ep-i-kīl'-i-um)
Epicrates (e-pik'-ra-tēz)
Epidendrum* (ep-id-en'-drum)
epididymis (ep-i-did'-i-mis)
Epigaea* (ep-i-jē'-a̧)
epigamic (ep-i-gam'-ik)
epigenesis (ep-i-jen'-e-sis)
epigeous (ep-ij-ē'-us) of the earth.
epigynous (ep-ij'-i-nus)
Epihippus (ep-i-hip'-us)

Epibulus (ep-ib′-u-lus)
epilimnion (ep-i-lim-nī′-on, ep-i-lim′-ni-on)
Epilobium* (ep-i-lō′-bi-um, ep-il-ob′-i-um)
Epimachus (e-pim′-a̯-kus, ep-im′-a̯-kus)
Epimartyria (ep-i-mâr-ti′-ri-a̯)
Epimedium* (ep-im-ē′-di-um)
epimere (ep′-i-mēr)
epimerite (ep-i-mēr′-īt, ep′-i-mêr-īt)
epimeron (ep-i-mē′-ron)
epiotic (ep-i-ōt′-ik)
Epipactis* (ep-i-pak′-tis)
Epiphyllum* (ep-if-il′-um)
epiphysis (e-pif′-i-sis, pl. e-pif′-i-sēz)
epiploön (e-pip′-lŏ-on)
epipodite (ep-ip′-ŏ-dīt)
epipodium (ep-i-pō′-di-um)
Epipogium (ep-i-pō′-ji-um)
epithelium (ep-i-thē′-li-um, pl. ep-i-thē′-li-a̯)
epithymoides (ep-ith-ī-mo-ī′-dēz) thyme-like.
epitoke (ep′-i-tōk)
epitokus (ep-it′-ŏ-kus)
Epitonium (ep-i-tōn′-i-um)
Epochra (ep-ok′-ra̯)
epsilus (ep-sī′-lus) somewhat bare.
Eptatretus (ep-ta̯-trē′-tus)
Eptesicus (ep-tes′-i-kus)
equine (ek′-wīn)
Equisetum* (ek-wi-sē′-tum)
Equus (ek′-wūs)
Eragrostis* (er-a̯-gros′-tis)
Eranthemum* (ē-ran′-the-mum)
Eranthis* (ē-ran′-this)

Erato (er'-a-tō)
Erax (ē'-raks)
erebenus (er-e'-ben-us) black.
Erechtites* (er-ek-tī'-tēz)
erector (er-ek'-tôr)
erectus (ē-rekt'-us) straight up.
Eremian (er-ē'-mi-an)
eremicola (er-ē-mik'-ol-ą) a desert-dweller.
eremicolor (er-ēm-i-kul'-ôr)
eremicus (er-ē'-mik-us) of deserts, pertaining to
    deserts or sandy plains, solitary, lonely.
eremobic (er-ē-mō'-bik)
eremology (er-ēm-ol'-ô-ji)
Eremomela (er-ê-mom'-e-lą)
Eremophila (er-ē-mof'-i-lą)
eremophyte (er-ēm'-of-īt)
Eremopterix (er-ēm-op'-ter-iks)
Eremorhax (er-ē'-mô-racks)
Eremurus* (er-ē-mū'-rus)
erepsin (er-ep'-sin)
Erethizon (er-e-thī'-zōn)
Eretmochelys (er-et-mok'-e-lis)
Ereunetes (e-rū-nē'-tēz)
ergates (er-gā'-tēz) a worker.
Ergaticus (er-gat'-i-kus)
Erica* (e-rī'-ką)
Ericameria* (e-rī-ką-me'-ri-ą)
ericetorum (e-rī-sē-tō'-rum) of heather, heather-
    loving.
ericifolius (er-īs-i-fol'-i-us, er-īs-i-fō'-li-us) erica-
    leaved.
Erigenia* (ē-ri-jen'-i-ą)

Erethizon <Gr. *erethizon*, the porcupine. Pronounced: er-e-thī′-zon, not
er-eth′-i-zōn.

erigens (ē′-ri-jenz) raising.
Erigeron* (ē-rij′-er-ōn, ē-rīj′-er-on)
Erignathus (e-rig′-na-thus)
Erigone (ê-rig′-ô-nē)
erinaceus (er-in-ā′-se-us)
Erineum (er-ī′-ne-um)
Eringium* (er-in′-ji-um)
Erinus* (er-ī′-nus)
Eriobotrya* (er-i-ob-ot′-ri-ạ)
Eriocaulon* (er-i-ok-ô′-lon)
Eriocera (er-i-os′-e-rạ)
Eriochilus* (er-i-ok-ī′-lus)
Eriochloa* (er-i-ok′-lŏ-ạ)
Eriocoma* (er-i-ok′-om-ạ)
Eriodes (er-i-ō′-dēz)
Eriogonum* (er-i-og′-ôn-um)
eriomerus (er-i-o′-me-rus) with woolly parts.
Eriophorum* (er-i-of′-ôr-um)
eriophorus (er-i-of′-ôr-us) wool-bearing.
Eriophyes (er-i-ô-fī′-ēz)

Eriogonum <Gr. *erios*, wool+*gony*, joint. Accent on antepenult (og) since the penult ŏ is not considered long it being derived from the Gr. short *o*, (omicron). Pronounced: er-i-og'-ŏ-num.

eriophylla (er-i-of-il'-ạ) woolly-leaved.

Eriophyllum* (er-i-of-il'-um)

eriospathus (er-i-os-pā'-thus) woolly-spathed.

Eriphia (e-rif'-i-ạ)

Erismatura (er-is-mat-ūr'-ạ)

Eristalis (er-is'-tạ-lis)

Erisyphe* (er-is-ī'-fē)

Erithrina* (er-ith-rī'-nạ)

ermineus (êr-min'-e-us) ermine-like, spotted like the ermine.

Erodium* (ẹ-rōd'-i-um)

Eriophyllum <Gr. *erios*, wool+*pyllon*, leaf. The penult is long because the vowel is followed by a double consonant. Pronounced: e-ri-of-il'-um, not er-i-of'-il-um.

erogenous (ē-roj'-e-nus)
Erophila* (er-of'-il-ạ)
erosus (ē-rō'-sus) jagged, gnawed.
Erotylidae (er-ộ-til'-i-dē)
erraticus (er-āt'-ik-us) wandering, straying.
erromenus (er-om'-en-us) strong, robust.
erubescens (ē-rub-es'-senz) blushing.
erucifolius (ē-rū-si-fol'-i-us, ê-rū-si-fō'-li-us) with
  Eruca-like leaves.
eruciform (ê-rū'-si-fôrm)
erumpens (ē-rum'-penz) breaking forth, bursting.
Ervum* (êr'-vum)
Eryngium* (ē-rin'-ji-um)
Erysimum* (e-ris'-im-um)
Erythacus (er-ith'-à-kus)
Erythea* (er-ith-ē'-ạ)
erythraeus (er-ith-rē'-us) reddish.
Erythrea* (er-ith-rē'-ạ)
Erythrina (er-ith-rī'-nạ)
erythrocyte (er-ith'-rō-sīt)
Erythronium* (er-ith-rō'-ni-um)
erythropus (er-ith'-rop-us) red-footed, red stalked.
Eryx (ē'-riks)
Escallonia* (es-ka-lō'-ni-ạ)
Eschara (es'-kä-rạ)
Eschscholtzia* (es-sholt'-zi-ạ)
esculentus (es-kul-ent'-us) edible.
esodic (ê-sod'-ik)
esophagus (ê-so'-fà-gus)
Esox (ē'-soks)
esoteric (es-ō-ter'-ik)
Estigmene (es-tig-mē'-nē)

Ethmia (eth'-mi-ą)
etiolation (ē-ti-ȯ-lā'-shun)
Euarctos (ū-ârk'-tos)
Eublepharis (ū-blef'-à-ris)
Eucharis (ū'-ka-ris)
Euchira (ů-kī'-rą)
Euchlaena* (ū-klē'-ną)
euchlorus (ū-klō'-rus) beautiful-green.
Euchone (ū-kō'-nē)
Euchoreutes (ū-kôr-o͞o'-tēz)
Euchroma* (ū-krō'-mą)
Euclea (ū-klē'-ą)
Eucleidae* (ū-klē'-i-dē)
Eucnetus (ūk-nē'-tus)
Eucnide* (ū-knī'-dē)
Eucodonia* (ū-kō-dō'-ni-ą)
Euconnus (ū-kon'-us)
Eucope (ū-kō'-pē)
Eudistylia (ū-di-stī'-li-ą)
Eudocimus (ū-dos'-i-mus)
Eudolon* (ū'-dol-on)
Eudynamis (ū-dī'-nà-mis)
Euelephus (ū-el'-e-fus)
Eufragia* (ū-frā'-ji-ą)
Eugenes (ū-jē'-nēz)
eugenics (ū-jen'-iks)
Euglandina (ū-glan-dī'-ną)
Eulabes (ū'-lå-bēz)
Eulalia* (ū-lal'-i-ą)
Eulecanium (ū-lek-ān'-i-um)
Eulophidae (ū-lof'-i-dē)
Eulophus* (ū'-lō-fus)

Eumeces (ū-mē′-sēz)
Eumenes (ū′-men-ēz)
Eumenidae (ū-men′-i-dē)
Eumycophyta (ū-mī-kof′-it-ạ)
Eunice (ū-nī′-sē)
Euonymus* (ū-ō′-nim-us, ū-on′-i-mus)
Eupagurus (ū-pa-gū′-rus)
Eupatorium* (ū-pȧ-tō′-ri-um, ū-pat-ôr′-i-um)
Eupetes (ū′-pe-tēz)
Euphagus (ū′-fag-us)
Euphausia (ū-fä-ūsh′-i-ạ)
Euphlebia* (ū-fleb′-i-ạ)
Euphorbia* (ū-fôr′-bi-ạ)
Euphrasia* (ū-frā′-shi-ạ, ū-frā′-si-ạ)
Euphuta (ū-fū′-tạ)
Euplectella (ū-plek-tel′-ạ)
Eupleres (ū-plē′-rēz)
Euplotes (ū-plō′-tēz)
Eupoda (ū-pō′-dạ, ū′-pod-ạ)
Eupodotis (ū-pȯ-dō′-tis)
Euproctis (ū-prok′-tis)
Eupsalis (ūp′-sal-is)
Euptelea* (ūp-tē′-lê-ạ)
eurocarpus (ū-rȯ-kâr′-pus) with broad fruit.
europhilus (ū-rof′-il-us) loving the southeast wind.
Euryalae (ū-ri-āl′-ē)
Eurycea (ū-ris′-ê-ạ)
euryhaline (ū-ri-hal′-īn)
Eurymus (ū′-ri-mus)
euryphagus (ū-rif′-ȧ-gus)
Eurytoma (ū-rit′-ȏm-ạ)
eurytopic (ū-ri-top′-ik)

Euscaphis* (ū'-skaf-is)
Euschistus (ū-shis'-tus)
Eustachian (ū-stā'-ki-an)
eustele (ūs-tē'-lē)
Eustoma* (ū'-stom-a̤)
Eutaenia (ū-tē'-ni-a̤)
Eutamias (ū-tā'-mi-as)
euthenics (ū-then'-iks)
euthycomous (ū-thik'-ôm-us)
Eutrema* (ū-trē'-ma̤)
Euxesta (ūks-es'-ta̤)
evagor (ē-vā'-gôr) wandering, roaming; also, ful-
    filling.
evanidus (ē-vā'-ni-dus) frail, feeble.
Evaniidae (ē-van-ī'-i-dē)
Evax* (ē'-vaks)
evexus (ē-veks'-us) rounding off near the top.
Evides (ev'-i-dēz)
evocator (ev-ok-āt'-or)
evolution (e-vō-lū'-shun; in England, ē-vō-lū'-
    shun)
Evolvulus* (ē-vol'-vul-us)
evotis (ē-vō'-tis)
Evotomys (ē-vōt'-ô-mis)
Exacum (eks'-ak-um)
Excaecaria (eks-sē-kā'-ri-a̤)
excelsior (ek-sel'-si-ôr) still higher.
exciple (ek'-si-pl)
excisus (ek-sī'-sus) cut away.
exconjugant (eks-kon'-jo͞o-gant)
excrement (eks'-krê-ment)
excreta (eks-krē'-ta̤)

excretive (eks-krē′-tiv, eks′-krĕ-tiv)
excretory (eks′-krē-tō-ri)
Exetastes* (eks-e-tas′-tēz)
exhale (eks-hāl′, eg-zāl′)
exiguus (eks-ij′-u-us) briefly, sparingly, small.
exilis (ex-īl′-is) small, weak, tender.
eximius (eks-ī′-mi-us) select, uncommon.
Exitelia* (eks-it-ē′-li-ạ)
exites (eks′-īts)
Exogonium (eks-ȯ-gōn′-i-um)
exogyrus (eks-oj′-i-rus)
exopodite (eks-op′-ȯ-dīt)
Exoprosopa (eks-op-ros-ōp′-ạ)
Exothea* (eks-oth′-ê-ạ)
Exothostemon (eks-ȯ-thos′-tê-mon)
exotic (egz-ot′-ik)
exoticus (eks-ot′-ik-us) from another country.
exsputus (eks-spū′-tus) spit out, removed.
extensus (eks-ten′-sus) spread out, stretched out.
extimus (eks′-ti-mus) most remote.
exustus (eks-us′-tus) burned up, consumed.
exuviae (eks-ū′-vi-ē)

F

faba (fab′-ạ) a bean.
Fabaceae* (fab-ā′-sê-ē)
Fabago* (fab-ā′-gō)
fabarius (fab-ā′-ri-us) bean-like.
Fabia (fā′-bi-ạ)
facies (fas′-i-ēz, fā′-shi-ēz) face, figure, shape.
faeces (fē′-sēz, pl. of L. *fex*)
Fagara* (fā-gā′-rạ)

fagineus (fā-jin′-e-us) of beech, of the beech tree.
Fagonia* (fȧ-gō′-ni-ạ)
Fagopyrum* (fȧ-gŏ-pī′-rum, fȧ-gop-ī′-rum)
Fagus* (fā′-gus)
falcatus (fal-kā′-tus) hooked, curved, sickle-shaped.
falcinellus (fal-sin-e′-lus) a small scythe.
falciparus (fal-si′-par-us) sickle-producing.

Falcon <early modern English
<Middle English *fawken* or *falkon*
<Late Gr. *falkon*. Pronounced:
fô′-kn.

—MVD

falcon (fô′-kn, fôl′-kn)
falconet (fô′-kŏ-net, fal′-kŏ-net)
falconry (fô′-kn-ri)
fallax (fal′-aks) deceptive.
Farancia (fa-ran′-shi-ạ)
farinose (far′-i-nōs)
farinosus (far-ī-nō′-sus) of meal, mealy.
fascia (fa′-shi-ạ, pl. fa′-shi-ē)
fasciatus (fas-si-ā′-tus) of bundles, bundled.
fascinator (fas-sin-ā′-tôr) an enchanter.
fasciola (fas-si′-o-lạ) a strip of cloth.

fasciolar (fas-si'-o-lâr)

Fasciolaria (fas-si-o-lā'-ri-ạ)

fasciole (fas'-si-ōl)

fastigiatus (fas-ti-ji-ā'-tus)

fastigium (fas-ti'-ji-um)

fastuosus (fas-tu-ō'-sus) proud.

Fatsia* (fat'-si-ạ)

fatuus (fat'-u-us) insipid, tasteless; also, simple, foolish.

fauces (fô'-sēz, sing. fô'-ks)

faustus (fôs'-tus) favorable, fortunate, auspicious.

faveolus (fav-ê'-ôl-us)

favulosus (fav-u-lōs'-us) full of small cells, a honey comb.

febrile (fē'-bril, feb'-ril)

fecal (fē'-kl)

feces (fē'-sēz)

fecund (fē'-kund, fek'-und)

fecundity (fê-kun'-di-ti)

Fedia* (fē'-di-ạ)

Feijoa* (fê-jō'-ạ)

Felicia* (fê-li'-shi-ạ)

feline (fē'-līn, fē'-lin)

Felis (fē'-lis)

fenisex (fē'-ni-seks) a mower.

fennec (fen'-ek)

feral (fē'-ral)

ferreus (fer'-e-us) made of iron; also, firm, fixed.

ferrugineus (fer-ū-ji'-ne-us) dark-red, rust-colored, dusky.

Ferula* (fer'-ul-ạ)

festinus (fes-tīn'-us) quick.

fetosus (fē-tōs'-us) prolific.
Festuca* (fes-tū'-kạ)
fetid (fē'-tid, fet'-id)
fibril (fī'-bril)
fibulare (fib-ŭ-lā'-rē)
Ficaria* (fī-kā'-ri-ạ)
ficifolius (fī-si-fol'-i-us, fī-si-fō'-li-us) with leaves like the fig tree (*Ficus*).
ficiform (fīs'-i-fôrm) fig-form
Ficimia (fī-sim'-i-ạ)
Ficus* (fī'-kus)
Fidonia* (fī-dō'-ni-ạ)
figwort (fig'-wêrt)
Filago* (fi-lā'-gō)
Filaria (fil-ā'-ri-ạ)
Filicinae* (fil-i-sī'-nē)
filicoides (fil-ik-o-ī'-dēz)
filicula (fil-ik'-ul-ạ) a rock-fern, polypody.
filiferus (fī-lif'-er-us) bearing threads.
filiformis (fī-li-fôr'-mis) thread-like in form.
Filipendula* (fī-lip-en'-du-lạ)
Filipes* (fī'-lip-ēz)
Filistata (fil-is-tā'-tạ)
filoplume (fī'-lō-plūm)
filosa (fīl-ōs'-ạ) full of threads.
filose (fī'-lōs)
fimbriatus (fim-bri-ā'-tus) fringed, cut in shreds.
Firmiana* (fêr-mi-ā'-nạ)
fission (fi'-shun)
fissiparous (fis-ip'-ar-us)
fissus (fis'-us) divided, separated.
fistulosus (fis-tu-lō'-sus) tubular, pipe-like.

flabella (flā-bel′-um)
Flabellina (flā-bel-īn′-ạ)
flaccid (flak′-sid)
flaccus (flak′-us) flabby, hanging down.
Flacourtia* (flak-ôrt′-i-ạ)
flagellaris (fla-jel-ā′-ris) whip-like.
flagellum (fla-jel′-um, pl. fla-jel′-ạ)
flammeolus (fla-me′-ol-us)
flammeus (fla′-me-us) flaming, fiery-red.
flammulatus (flam-ul-ā′-tus) provided with little
    flames.
flavescens (flā-ves′-senz) growing yellow, yellow.
flavidus (flā′-vi-dus) of golden yellow, yellowish.
flavirameus (flāv-i-rā′-me-us) yellow-branched.
flavovirens (flā-vō′-vi-renz) yellow-green.
flavus (flā′-vus) golden-yellow, of the color of flax.
flexuosus (fleks-u-ō′-sus) full of turns or windings,
    tortuous, crooked.
Floerkia* (flēr′-ki-ạ)
flora (flō′-rạ)
flore-pleno (flō′-rē-plē′-nō) with full or double
    flowers.
floridanus (flôr-id-ā′-nus) of Florida.
Floscularia (flos-kū-la′-ri-ạ)
Flourensia* (flūr-en′-si-ạ)
fluitans (flu′-i-tans, floo′-it-anz) floating.
fluviatilis (flu-vi-ā′-ti-lis) of or belonging to a
    river.
fodiens (fod′-i-enz) digging.
Foeniculum* (fē-nik′-ul-um)
foetens (fē′-tenz) ill-scented, stinking.
foetid (fē′-tid, fe′-tid)

foetidissimus (fē-tid-is'-i-mus) most fetid, foul-odored.

foetidus (fē'-ti-dus) ill smelling, foul, stinking.

foetus (fē'-tus, pl. fē'-tus-ez)

foliation (fō-li-ā'-shun)

foliole (fō'-li-ōl)

foliolosus (fol-i-ol-ō'-sus) with leaflets.

folium (fō'-li-um)

folsomi (fōl'-som-ī)

fontanus (fon-tān'-us) of or from a spring or fountain.

fontinalis (fon-tin-ā'-lis) relating to a spring.

foramen (fō-rā'-men, pl. fō-rām'-in-ạ)

Foraminifera (fō-ram-i-nif'-êr-ạ)

Forchammeria* (fôr-sham-ē'-ri-ạ)

forehead (fôr'-ed)

Forestiera* (fôr-es-ti-ē'-rạ)

forficatus (fôr-fik-ā'-tus) deeply notched.

forficulidae (fôr-fi-kū'-li-dē)

Formica (fôr-mī'-kạ)

Formicidae (fôr-mis'-i-dē)

formosus (fôr-mō'-sus) beautiful, finely formed.

fossa (fos'-ạ, pl. fos'-ē)

fossor (fo'-sôr) a digger.

Fouquieria* (fū-ki-ē'-ri-ạ)

fovea (fō'-vê-ạ)

foveiform (fō-vē'-i-form)

foveola (fō-vē'-ô-lạ)

foveolate (fō-vē'-ô-lāt)

Fragaria* (frā-gā'-ri-ạ)

fragiferus (frā-ji'-fer-us) strawberry-bearing.

fragilis (fra'-ji-lis) fragile, brittle; also, weak.

fragrosus (frag-rō'-sus) fragile.

Francolinus (frang-kô-lī'-nus)

Frankenia* (frank-ēn'-i-ạ)

frater (frā'-ter) a brother.

Fratercula (frā-têr'-kǔ-lạ)

fraterculus (frā-ter'-kù-lus) a little brother; also, of unknown parents.

Fraxinus* (frak'-si-nus)

Fregata (frē-gā'-tạ)

Fregilus (frej'-i-lus)

frenulatus (frē-nu-lā'-tus) bridled.

frenulum (fren'-u-lum, frē'-nu-lum)

frenum (frē'-num)

frequens (fre'-kwenz) often, repeatedly.

Friesia* (frē'-si-ạ)

frigidus (frī'-ji-dus) cold.

Fritillaria* (frit-il-ā'-ri-ạ)

frondator (fron-dā'-tor) one who strips leaves, a pruner.

frons (fronz) a leafy branch.

frontal (frun'-tal)

fructivorous (fruk-ti'-vôr-us) fruit-eating.

fructose (fruk'-tōs, frook'-tos)

frumentaceous (froo-men-tā'-shus)

frustror (frus'-trôr) deceiving, useless.

frustulentus (frust-u-len'-tus) filled with small pieces, a bit, a piece.

frutescens (frut-es'-enz) becoming shrubby.

frutex (frut'-eks) a bush.

fruticosus (frut-i-kō'-sus) shrubby, bushy, full of bushes.

fruticulosus (frut-ik-ul-ō'-sus) putting forth many small shoots, to put forth shoots.

fuchsia (fū'-shi-ạ)

Fuchsia* (fŏŏk'-si-ạ, fū'-shi-ạ)

fucosus (fū-kō'-sus)

Fucus* (fū'-kus)

fugacious (fū-gā'-shus)

fugax (fug'-aks) swift, fleet.

fugiens (fu'-ji-enz) fleeing.

fulgens (ful'-jenz) glowing.

fulgidus (ful'-ji-dus) glittering, flashing.

Fulgoridae (ful-gôr'-i-dē)

fulgurans (ful'-gu-ranz) flashing, glittering.

Fulica (fū'-li-kạ)

fulicarius (fū-lik-ā'-ri-us) coot-like.

Fulix (fū'-liks)

fullonum (ful-ōn'-um) of one who fulls cloth.

fulmar (fŏŏl-mêr)

Fulmarus (fŏŏl'-mȧ-rus)

fulvus (ful'-vus) tawny, gold-colored, deep yellow.

Fumaria* (fū-mā'-ri-ạ)

fumeus (fū'-me-us) smoky, full of smoke.

Funastrum* (fū-nas'-trum)

funebralis (fū-ne-brā'-lis) pertaining to the dead.

Fungi* (fun'-jī)

Fungia (fun'-ji-ạ)

fungus (fung'-us, pl. fun'-jī)

funicle (fūn'-ikl)

funiculus (fūn-ik'-ūl-us)

furax (fū'-raks) given to stealing.

furcatus (fûr-kā'-tus) forked.

Furcraea* (fûr-krē'-ạ)

furcula (fûr'-kŭ-lạ)
furfurosus (fûr-fûr-ō'-sus) brownish, like bran.
furvus (fûr'-vus) dark, dusky, black.
fuscatus (fus-kā'-tus)
fuscipes (fus'-si-pēz) dusky-footed or black-footed.
fuscus (fus'-kus) dark-tawny.
Fusicladium* (fū-sik-lad'-i-um)
fusiform (fū'-si-fôrm)
Fusinus (fū'-sin-us)

G

Galactia* (ga-lak'-ti-ạ, ga-lak'-shi-ạ)
galactophorous (gal-akt-of'-ôr-us)
Galago (ga-lā'-gō)
Galanthus* (ga-lan'-thus)
Galax* (gā'-laks, gal'-aks)
galea (gal'-e-ạ, gā'-lē-ạ) a helmet.
Galeata* (gal-e-ā'-tạ)
galeatus (gal-e-ā'-tus) helmeted.
Galedupa* (gal-ē'-dup-ạ)
Galega* (gal-ē'-gạ)
Galemys (gal'-e-mis)
Galeobdolon (gā-lē-ob'-dol-on, gal-e-ob'-dol-on)
galeodes (gal-e-ō'-dēz) like a shark.
Galeopithecus (gā-le-ō-pi-thē'-kus)
Galeopsis* (gā-le-op'-sis, gal-e-op'-sis)
Galera (gal-ē'-rạ)
galericulatus (gal-ē-ri-kul-ā'-tus) hooded.
Galerida (gal-er'-id-ạ)
Galeruca (gal-ê-rōō'-kạ)
Galgulus (gal'-gu-lus)

Galictis (gal-ik'-tis)
Galidia (gā-lid'-i-ą)
Galium* (gā'-li-um, gal'-i-um)
Galleria (gal-er'-i-ą)
Galleriidae (gal-er-ī'-i-dē)
Gallerucella (gal-er-ūs-el'-ą)
gallicus (gal'-i-kus) French, from Gaul.
gallina (gal-īn'-ą) a hen.
Gallinago (gal-i-nā'-gō)
Gallinula (gal-in'-ū-lą)
Gallirallus (gal-i-ral'-us)
Gallus (gal'-us)
gambusia (gam-bū'-si-ą) nothing.
gametal (gam-ē'-tal)
gametangium (gam-ê-tan'-ji-um)
gamete (ga'-mēt, ga-mēt')
gametids (gam-ē'-tidz)
gametogenesis (gam-ê-tŏ-jen'-e-sis)
gametophyta (ga-mê-tof'-it-ą)
gametophyte (ga-mē'-tŏ-fīt)
Gammaridia (gam-âr-id'-i-ą)
Gammarus (gam'-ȧ-rus)
Gamolepis* (gam-ol'-ep-is)
gangrenosus (gan-gren-ō'-sus) full of eating sores.
gape (gap, gāp)
Garcinia* (gâr-sin'-i-ą)
Gardenia* (gâr-dē'-ni-ą; gâr-den'-i-ą)
Gasteria* (gas-tē'-ri-ą, gas-ter'-i-ą)
gastraea (gas-trē'-ą)
Gastridium* (gas-trid'-i-um)
Gastrochaena (gas-trŏ-kē'-ną
gastrocnemius (gas-trok-nē'-mi-us)

Gastropacha (gas-trop'-á-ką)
Gastrophilus (gas-trof'-i-lus)
Gastrophryne (gas-trŏ-frī'-nē)
Gastropoda (gas-trop'-ŏd-ą)
gastrula (gas'-trū-lą)
Gaultheria* (gŏl-thē'-ri-ą)
Gaura* (gŏ'-rą)
gausapatus (gŏ-sa-pā'-tus) covered over, covered
with felt.
Gavia (gā'-vi-ą)
gavial(gā'-vi-al)
Gavialis (gā-vi-ā'-lis)
Gayophytum* (gā-ŏ'-fit-um)
Gazania* (gā-zā'-ni-ą)
Geaster* (jē'-as-têr)
Geatractus (jē-at-rak'-tus)
Gecarcinus (jê-kâr'-si-nus)
geebung (jē'-bung)
Geissorhiza (gī-sŏ-rī'-zą)
gelasinatus (jel-as-in-ā'-tus) with dimples.
gelasinus (jel-as'-i-nus)
Gelastocoridae (jē-las-tŏ-kôr'-i-dē)
Gelastocoris (jē-las-tok'-ôr-is)
Gelechia (jē-lē'-ki-ą)
Gelechiidae (jē-lēk-ī'-i-de)
Gelidium* (jê-lid'-i-um)
gelidus (jel'-i-dus) icy cold, frosty.
Gelochelidon (jel-ŏ-kel-ī'-dōn)
Gelsemium* (jel-sē'-mi-um)
gemmiparus (jem-ip'-ar-us)
gemmule (jem'-ūl)
gena (jē'-ną)

genealogy (jen-ê-al'-ô-ji, jē-nê-al'-ô-ji)
generalis (jen-er-ā'-lis) general, prevailing.
Generium* (jen-er'-i-um)
generosus (jen-er-ōs'-us) of noble birth, eminent, superior, excellent.
Genetta (jê-net'-ą)
genic (jen'-ik)
geniculatus (jen-ik-ul-ā'-tus) with bended knee, bent, curved.
geniculum (jen-ik'-ul-um)
Genipa* (jen-ī'-pą)
Genista* (jen-is'-tą)
genital (jen'-i-tal)
Gennaeus (jen-ē'-us)
genotype (jen'-ô-tīp)
Gentiana* (jen-shi-ā'-ną)
gentilis (jen-tī'-lis) belonging to the same stock; also, foreign.
genys (jen'-is)
geobionts (jē-ôb-ī'-onts)
Geococcyx (jē-ô-kok'-siks)
Geocoris (jē-ok'-ôr-is)
Geogale (jē-og'-a-lē)
Geometridae (jē-ô-met'-ri-dē)
Geomys (jē'-ô-mis)
Geonoma* (jē-on'-ô-ma, je-ô'-no-mą)
geophilus (jē-of'-il-us) ground-loving.
Geophis (jē'-of-is)
Georyssus (jē-ô-ris'-us)
Geothlypis (jē-oth'-lip-is)
geotonus (jē-ot'-ô-nus)
Geotripes (jē-ô-trī'-pēz)

geotropism (jē-ot'-rŏ-pizm)
Geotrygon (jē-ŏ-trī'-gon)
gephyrea (je-fī'-rē-ạ, je-fi-rē'-ạ)
gephyrocercal (jef-ir-ŏ-sêr'-kal)
Geraea* (je-rē'-ạ)
Geranium* (jer-ā'-ni-um)
Gerbera* (gêr'-bêr-ạ, ger-bē'-rạ)
Gerbillus (jêr-bil'-us)
germigen (jêr'-mi-jen)
Geropogon* (jer-op-ō'-gōn)
Gerrhonotus (jer-ŏ-nō'-tus)
Gerridae (jer'-i-dē)
gestalt (ge-stält')
getulus (jē-tū'-lus) of Lybia, of the African coast.
Geum* (jē'-um)
Gibberella* (jib-êr-el'-ạ)
gibbifrons (gib'-i-fronz) with swollen front.
gibbosus (gib-ō'-sus) full of humps, badly hump-backed.
gibbus (gib'us) crooked, humped.
giganteus (jī-gan'-te-us) very large.
gigas (jī'-gas) a giant.
Gilia (jil'-i-ạ, gil'-i-ạ)
Gillenia* (gil-ē'-ni-ạ, jil-ē'-ni-à)
gilvus (gil'-vus) pale-yellow.
gingival (jin-jī'-val, jin'-jiv-al)
Gingko* (ging'-kō, jing'-kō)
Gingla (jin'-glạ)
ginglymus (jing'-li-mus, ging'-li-mus)
Ginkgo* (gin'-kō, jing-kō)
Giraffa (jī-ra'-fạ)
Githago* (gith-ā'-gō)

glabellus (glab-el'-us) smoothish.

glaber (gla'-bêr) without hair, bald, smooth.

glabriusculus (glab-ri-us'-ku-lus) somewhat bald, nearly without hair.

Gladiolus* (glad-ī'-ô-lus, glad-i-ō'-lus)

glanduliferus (glan-dul-if'-er-us) gland-bearing, glandular.

glandulosus (glan-dul-ō'-sus) full of kernels, full of glands.

Glareola (gla-rē'-ô-lą)

Glaucidium (glô-sid'-i-um)

glaucinus (glô'-sin-us) blue-gray, silvery, gray.

Glaucionetta (glô-si-ô-net'-ą)

Glaucium* (glô'-si-um)

Glaucomys (glô'-kô-mis)

glaucopsis (glô-kop'-sis) glaucous-like.

glaucus (glô'-kus) sea-green, covered with a "bloom."

Glaux* (glôks)

gleba (glē'-ba)

glebula (glē'-bul-ą)

glinus (glī'-nus)

gliosomes (glī'-ôs-ōmz)

Glires (glī'-rēz)

gliriform (glī'-ri-fôrm)

glischrus (glis'-krus) sticky, clammy.

globator (glob-ā'-tôr) maker of a globe.

Globicephalus (glō-bi-sef'-al-us)

Globigerina (glō-bi-je-rī'-ną)

Globiocephalus (glō-bi-ô-sef'-al-us)

globosus (glob-ō'-sus) round, spherical.

Globularia* (glob-u-lā'-ri-ą)

globule (glob'-ūl)
globuliferus (glob-ul-if'-êr-us) bearing globules.
globulin (glob'-ū-lin)
globus (glob'-us, pl. glob'-ī)
glochid (glō'-kid)
glochidium (glō-kid'-i-um)
Gloeocapsa* (glē-ō-kap'-sạ)
glomeratus (glom-er-ā'-tus) gathered into a round
    mass.
glomerulus (glom-er'-u-lus)
Glossina (glōs-ī'-na, glos-īn'-ạ)
Glossocomia* (glōs-ok-om'-i-ạ, glos-ok-om'-i-ạ)
Glossopetalon* (glōs-ō-pet'-al-on,   glos-ō-pet'-al-
    on)
Glossophaga (glōs-of'-ȧ-gạ, glos-of'-ȧ-gạ)
Glottidia (glō-tid'-i-ạ, glot-id'-i-ạ)
Glottiphyllum* (glō-ti-fi'-lum, glot-i-fi'-lum)
glumaceous (glū-mā'-shus)
gluteal (glū-tē'-al, glū'-tē-al)
glutinosus (glū-tin-ō'-sus) full of glue, tenacious.
Glyceria* (gli-se'-ri-ạ)
glycogen (glī'-kō-jen)
Glycymeris (glis-im'-e-ris)
glycyphyllus (glis-if-il'-us) with sweet leaves.
Glycyrrhiza (glis-i-rī'-zạ)
Glyptopleura* (glip-tō-plū'-rạ)
glyptospermus (glip-tō-spêr'-mus) with sculptured
    seed.
gnamptorhynchus (namp-tō-ring'-kus)
Gnaphalium* (na-fā'-li-um, na-fal'-i-um)
Gnophaela (gnof-ē'-lạ)
Gnorimoschema (nôr-im-os-kē'-mạ)

Gnostum (nos'-tum)
gnu (nū)
Godetia* (gō-dē'-shi-ạ)
Gomphrena* (gom-frē'-nạ)
gonad (gon'-ad)
gonadotropic (gon-ad-ȯ-trop'-ik)
gonangium (gon-an'-ji-um)
gonapophyses (gon-i-pof'-is-ēz)
Gonatocerus (gō-nat-os'-er-us)
Gonatopus (gō-nat'-op-us)
gondolus (gon'-do-lus) boat-shaped.
gongylodes (gon-ji-lō'-dēz) turnip-like.
Gongylonema (gon-ji-lȯ-nē'-mạ)
gonidia (gon-id'-i-ạ)
gonion (gōn'-i-on)
Gonionemus (gon-i-ȯ-nē'-mus)
Gonolobus* (gō-nol'-ȯ-bus, gōn-ol'-ȯ-bus)
gonotheca (gon-ȯ-thē'-kạ)
Gonyaulax (gon-i-ôl'-aks)
gonys (gon'-is)
gooseberry (goōs'-ber-i, goōz'-ber-i)
Gopherus (gō'-fêr-us)
goral (gō'-ral)
Gorilla (gȯ-ril'-ạ)
Gorytez (gôr-ī'-tēz)
goshawk (gos'-hôk)
Gossipium* (gos-ip'-i-um)
gourd (gôrd, goōrd)
Goveniana* (gov-ē-ni-ā'-nạ)
Gracilariidae (gras-i-la-rī'-i-dē)
gracilentus (gras-il-en'-tus) very slender.
gracilis (gras'-il-is) delicate, slender.

graecizans (grē'-si-zanz) becoming widespread.
Grallae (gral'-ē)
Grallatores (gral-a-tō'-rēz)
Grallina (gral-ī'-na̧)
gramineus (grā-mi'-ne-us) pertaining to grass, grassy.
grammacus (gram'-a-kus) consisting of lines, streaked.
Granatellus (gran-at-el'-us)
Granatum* (grā-nā'-tum)
grandiflorus (gran-dif-lō'-rus) large-flowered.
grandifolius (gran-di-fol'-i-us, gran-di-fō'-li-us) large-leaved.
grandis (grand'-is) large, great, full, abundant.
granulatus (grā-nul-ā'-tus) bearing small tubercules, covered with small granulations.
granulocyte (gran'-ů-lô-sīt)
granulosus (grā-nul-ō'-sus) full of grains.
Grapsidae (grap'-si-dē)
Graptemys (grapt'-e-mis)
graptolite (grap'-tô-līt)
Graptophyllum* (grap-tof-il'-um)
Gratiola* (grâ-ti'-ol-a̧, grā'-ti-ol-a̧)
graveolens (grav-e'-o-lenz) strong-scented.
graveolent (grav-e'-o-lent)
gravis (grav'-is) heavy.
Gregarina (greg-à-rīn'-a̧)
Gregarinida (greg-à-rin'-id-a̧)
gregarious (gre-gā'-ri-us)
Grevillea* (grev-il'-e-a̧)
grex (greks) a swarm, a herd.
Grias* (grī'-as)

grisbox (grīs'-box)

grisescens (gris-es'-senz) becoming or tending toward grey.

griseus (gris'-e-us, gris'-ê-us) gray.

Grison (gris'-ȯn)

grosbeak (grōs'-bēk)

grossularia (gros-ul-ā'-ri-ạ) pertaining to a gooseberry.

grossus (gros'-us) large, thick.

Grus (grūs, grus)

Gryllidae (gril'-i-dē)

Grylloblattodea (gril-ȯ-blat-o-dē'-ạ)

Gryllotalpidae (gril-ȯ-talp'-i-dē)

Grypanian (gri-pā'-ni-an)

Guaiacum* (gwī'-ȧ-kum)

guanaco (gwä-nä'-kō)

guano (gwan'-ō)

Gubernetes (gū-bêr-nē'-tēz)

guenon (gē-non')

guereza (ger'-ê-zạ)

guillemot (gil'-e-mot)

Guiraca (gwi-rā'-kạ)

Gulo (gū'-lō)

gulosus (gul-ō'-sus) big-mouthed, gluttonous.

gummosus (gum-ōs'-us) gummy, made of gum.

gutta (gu'-tạ, pl. gu'-tē)

guttation (gu-tā'-shun)

guttatus (gut-ā'-tus) spotted.

Guzmania* (gūz-man'-i-ạ)

Gyalecta (jī-ȧ-lek'-tạ)

Gyalopion (jī-al-ōp'-i-on)

gyas (ji'as) giant with a hundred arms.

Gygis (jī'-jis)
Gyminda* (jim'-in-dạ)
Gymnadenia* (jim-na-dē'-ni-ạ)
gymnantherus (jim-nan'-thêr-us) naked-flowered.
gymnetrous (jim-nē'-trus)
Gymnocladus* (jim-nok'-la-dus)
Gymnogramme* (jim-nog-ram'-ē)
gymnoheliophilist (jim-nȯ-hēl-i-of'-il-ist)
Gymnolaemata (jim-nȯ-lē'-mȧ-tạ)
Gymnophiona (jim-nȯ-fī'-ȯ-nạ)
Gymnorhina (jim-nȯ-rī'-nạ)
gymnosperm (jim'-nȯ-spêrm)
gymnospermae (jim-nos-pêr'-mē)
gynandromorph (jin-an'-drȯ-môrf)
gynase (jin'-ās)
gyne (jī'-nē)
gynecology (jin-ē-kol'-ȯ-ji, jī-nȯ-kol'-ȯ-ji)
gynobase (jin'-ȯ-bās, jī'-nȯ-bās)
gynoecium (jin-ê'-shi-um, jin-ē'-si-um)
gynophore (jin'-ȯ-fôr, jī'-nȯ-fôr)
Gynura* (jin-ū'-rạ)
Gypaetus (jip-ā'-e-tus)
Gypohierax (jip-ȯ-hi'-êr-aks)
Gypona (jip'-on-ạ)
Gyps (jips)
Gypsophila* (jip-sof'-i-lạ)
Gypsophoca (jip-sof-ōk'-ạ)
gyration (jī-rā'-shun)
gyrfalcon (jêr'-fȯl-kun, jêr'-fȯl-kn)
Gyrinidae (ji-rin'-i-dē)
Gyrinophilus (ji-rin-of'-il-us)
Gyrinus (ji-rīn'-us, jī-rin'-us)

Gyrocerus (ji-ros'-ê-rus, jī-ros'-ê-rus)
Gyrocoryna (ji-rô-kō'-ri-na̧, jī-rô-kō'-ri-na̧)
Gyrocotyle (ji-rô-kot'-il, jī-rô-kot'-il)
Gyrodactylus (ji-rô-dak'-til-us, jī-rô-dak'-til-us)
gyroma (ji-rō'-ma̧, jī-rō'-ma̧)
Gyropidae (ji-rop'-i-dē, jī-rop'-i-dē)
gyrotoma (jī-rot'-ôm-a̧)
gyrus (jī'-rus, pl. jī'-rī)

H

Habenaria* (hab-ê-nā'-ri-a̧)
Habranthus* (hab-ran'-thus)
Hadena (ha̧-dē'-na̧)
Hadenoecus (had-e-nē'-kus)
Hadentomum (hā-den'-tô-mum)
Hadrosaurus (had-rô-sô'-rus)
haematin (hē'-ma̧-tin, hem'-a̧-tin)
Haematobia (hē-mat-ob'-ī-a̧)
haematology (hē-mat-ol'-ô-ji, hem-at-ol'-ôj-i)
Haematopinidae (hē-ma-tô-pin'-i-dē)
Haematopus (hē-mat'-ô-pus)
Haematoxylon* (hē-mat-oks'-il-on, hem-at-oks'-il-on)
haemoglobin (hē-mô-glō'-bin, hem-ô-glō'-bin)
haemophilia (hē-mô-fil'-i-a̧, hem-ô-fil'-i-a̧)
Hakea* (hā'-kê-a̧, hā'-ke-a̧)
Halcyon (hal'-si-ōn)
halepensis (hal-e-pen'-sis) of Aleppo.
Halesia (hāl'-zi-a̧, hāl-ē'-shi-a̧)
Haliaëtus (hal-i-ā'-e-tus)
halibut (hal'-i-but)
Halichoerus (hal-i-chē'-rus)

Haliaëtus <Gr. *haliaetos*, a
bird, prob. the osprey
<*hals*, the sea+*aetos*, the
eagle. Pronounced: hal-i-ā'-
e-tus.

Halicore (hal-ik'-ô-rē)
Halictidae (hal-ik'-ti-dē)
Halictus (hal-ik'-tus)
Halimium* (hal-im'-i-um)
halimus (hal'-i-mus) a plant, the orach.
Haliotis (hal-i-ō'-tis)
Haliplana (hal-ip'-lȧ-nạ)
Haliplidae (hal-ip'-li-dē)
halitus (hal'-i-tus)
Halmaturus (hal-ma-tū'-rus)
halophilous (hal-of'-il-us)
halophilus (hal-of'-il-us) salt-loving.
halophyte (hal'-ô-fīt)
halosere (hal'-ô-sēr)
Halosoma (hal-ô-sō'-mạ)
Halosydna (hal-os-id'-nạ)
Halsidota (hal-si-dō'-tạ)
halter (hal'-têr, pl. hal-tē'-rēz)
Haltica (hal'-ti-kạ)

Hamamelis* (ham-a-mē'-lis)
hamatum (hā-mā'-tum)
hamilifolius (ham-il-i-fol'-i-us, ham-il-i-fō'-li-us)
  with leaves like *Atriplex hamilus.*
hamulatus (hā-mu-lā'-tus) furnished with a small
  hook.
hamulus (hā'-mu-lus) a small hook.
hamus (hā'-mus) a hook.
Hapale (hap'-a-lē)
Haploa (hap-lō'-a)
haptera (hap'-tê-ra)
Harelda (ha-rel'-da)
Harenactis (ha-ren-ak'-tis)
harlequin (hâr'-lê-kwin)
Harpalus (hâr'-pal-us)
Harpephyllum* (hâr-pe-fil'-um)
harpes (hâr'-pēz)
hastaefolius (has-tē-fol'-i-us, has-tē-fō'-li-us) spear-
  leaved.
hastatus (has-tā'-tus) armed as with spears.
hastula (has'-tū-la)
haustor (hô'-stôr) a drawer of water.
Haworthia* (hô-wêrth'-thi-a)
Hebeandra* (hē-bē-an'-dra)
hebecarpus (hē-bē-kâr'-pus) pubescent-fruited.
hebes (heb'-ēs) blunt.
Hechtia* (hek'-ti-a)
Hedeoma* (hē-dē-ō'-ma, hed-ê-ō'-ma)
Hedera* (hed'-êr-a)
hederaceus (hed-er-ā'-se-us) of ivy, ivy-green.
hederaefolius (hed-er-ē-fol'-i-us, hed-er-ē-fō'-li-us)
  ivy-leaved.

hedonic (hē-don'-ik)
Hedychium* (hē-dik'-i-um)
Hedychrum (hē-di'-krum)
Hedymeles (hē-di-mēl'-ēz)
Hedyotis (hē-di-ō'-tis)
Hedysarum* (hē-dis'-à-rum)
Heisteria* (hīs-tē'-ri-ạ)
hekistotherm (hê-kist'-ô-thêrm)
Heladotherium (hel-à-dô-thē'-ri-um)
Helenium* (he-le'-ni-um, he-lē'-ni-um)
Heleocharis* (hel-ê-ok'-à-ris)
Heleodytes (hel-ê-ô-dī'-tēz)
Helianthemum* (hē-li-an'-thê-mum)
Helianthus* (hē-li-an'-thus)
Helice (hel'-i-sē)
Helichrysum* (hē-lik-rī'-sum)
Helicodiceros* (hel-ik-od-dis'-er-os)
helicoid (hel'-i-koid) coiled like a snail shell.
Helictis (hel-ik'-tis)
Helietta* (hel-i-et'-ạ)
Heliodinidae (hē-li-ô-din'-i-dē)
Heliophila (hē-li-of'-il-ạ)
Heliopsis* (hē-li-op'-sis)
Heliornis (hē-li-ôr'-nis)
Heliothis (hel-i-ōth'-is)
heliothropism (hē-li-ot'-rô-pizm)
Heliotropium* (hē-li-ot-rō-'pi-um)
Heliozela (hē-li-oz-ēl'-ạ)
Helisoma (hel-is-ōm'-ạ)
helix (hel'-iks, hē'-liks, pl. hel'-i-sēz, hē'-li-sēz)
Helleborus* (hel-eb'-ô-rus)
Helminthia* (hel-min'-thi-ạ)

Helmintherus (hel-min-thē′-rus)
helobius (hel-ō′-bi-us)
Heloderma (hē-lŏ-dêr′-mạ)
Helodromas (hel-ō′-dro-mas)
Helogale (hel-og′-ȧl-ē)
Helonias* (hel-ō′-ni-as)
Helorus (hel-ō′-rus)
Helosciadium* (hel-os-si-ad′-i-um)
Helostoma (hĕ-los′-tŏ-mạ)
helotism (hel′-ot-izm, hē′-lot-izm)
helveolus (hel-ve′-ol-us) pale yellow.
helvolus (hel′-vol-us) pale yellow.
Helxine* (hel-ksī′-nē)
hemal (hē′-mal)
Hemerobiidae (hem-er-ŏ-bī′-i-dē, hē-mer-ŏ-bī′-i-dē)
Hemerocallis* (hem-er-ŏ-kal′-is, hē-mer-ŏ-kal′-is)
Hemigale (hem-ig′-ȧ-lē)
Hemimeridae (hem-i-mer′-i-dē)
hemionus (hē-mī′-on-us) a half-ass, a mule.
Hemiphlebiidae (hem-i-fle-bī′-i-dē)
Hemiptera (hem-ip′-têr-ạ)
Hemisia (hem-is′-i-ạ)
Hemitragus (hem-it-rā′-gus)
hemocoel (hem′-ŏ-sēl)
hemoglobin (hē-mŏ-glō′-bin, hem-ŏ-glō′-bin)
Hepialidae (hē-pi-al′-i-dē)
heptalobus (hep-tȧ-lō′-bus) seven-lobed.
Heracleum* (her-a-klē′-um)
herbaceous (hêr-bā′-shus)
herbaceus (hêr-bā′-se-us) grassy, grass-green, with
    green stalks.
herbarium (hêr-bâr′-i-um, hêr-bā-ri′-um)

herbivorous (hêr-bi'-vō-rus)
Heriades (hêr-ī'-ad-ēz)
hermaphrodite (hêr-ma'-frȯ-dīt)
hermaphroditism (hêr-maf'-rȯ-dīt-izm)
Herminium* (hêr-min'-i-um)
Herniaria* (hêr-ni-ā'-ri-ạ)
Herodiones (hē-rō-di-ō'-nēz)
heron (he'-run)
Herpestes (hêr-pēs'-tēz, hêr-pes'-tēz)
Hesperaloe* (hes-per-al-ō'-ē)
Hesperiidae (hes-per-ī'-i-de)
Hesperiphona (hes-per-if-ōn'-ạ)
Hesperis* (hes'-per-is)
hesperius (hes-per'-i-us) of the West.
Hesperocallis* (hes-per-ȯ-kal'-is)
Hesperomys (hes-per'-ȯ-mis)
Hesperornis (hes-per-ôr'-nis)
Heteranthera* (het-er-an'-thē-rạ)
Heterocera (het-êr-os'-er-ạ)
heteroclitus (het-er-ȯ-klīt'-us)
Heterodon (het-er'-ȯ-don)
heterogeneity (het-er-ȯ-jē-nē'-i-ti)
heterogeny (het-er-oj'-e-ni)
Heterogeomys (het-er-ȯ-jē'-o-mis)
Heterolocha (het-er-ȯ-lōk'-ạ)
heteromerous (het-er-om'-êr-us)
Heteromys (het-er'-ȯ-mis)
heterophyllus (het-er-of-il'-us) with different
  leaves.
Heteroplectron (het-er-ȯ-plek'-tron)
heterosis (het-er-ō'-sis)
heterosporous (het-er-os'-pôr-us)

Heterotheca* (het-er-ô-thē′-kạ)
heterotropic (het-er-ô-trōp′-ik)
heterozygote (het-er-ô-zī′-gōt)
Heuchera* (hū-kē′-rạ, hū′-kê-rạ)
Hevea* (hē′-vê-ạ)
Hexactinellida (heks-ak-ti-nel′-i-dạ)
Hexagenia (heks-aj-ēn′-i-ạ)
Hexalectris* (heks-a-lek′-tris)
hexandrus (heks-an′-drus) having six anthers.
hians (hi′-anz) an opening, a gaping.
hiantulus (hi-an′-tu-lus) with a small opening.
hiatus (hī-ā′-tus)
hibernus (hī-bêr′-nus) belonging to winter.
Hibiscus* (hī-bis′-kus, hib-is′-kus)
hiemal (hī′-em-al)
hiemalis (hi-em-ā′-lis) belonging to winter.
hiemation (hī-em-ā′-shun)
Hieracium* (hī-êr-ā′-shi-um, hi-êr-ā′-shi-um)
Hierochloe* (hī-êr-ok′-lo-ē, hi-er-ok′-lo-ē)
Hilaria* (hi-lā′-ri-ạ)
hilum (hī′-lum)
hilus (hī′-lus)
Himantopus (hī-man′-tô-pus)
Himatione (him-at-i′-on-ē)
Hinnites (hi-nī′-tēz)
Hipparion (hi-pā′-ri-on)
Hippelates (hip-el-āt′-ēz)
Hippiscus (hip-is′-kus)
Hippoboscidae (hip-ô-bos′-i-dē)
hippocampus (hip-ô-kam′-pus)
Hippocrepis* (hip-ô-krē′-pis)
Hippodamia (hip-ô-dā-mi′-ạ)

Hippolestes (hip-ŏ-les′-tēz)
Hippolyte (hip-ol′-i-tē)
Hippophae* (hip-of′-à-ē)
Hippopotamus (hip-ŏ-pot′-à-mus)
Hippopus (hip′-ŏ-pus)
Hippuris* (hip-ū′-ris)
hircinus (hêr-si′-nus) of a goat, with smell like a goat.
hirsute (hêr′-sūt, hêr-sūt′)
hirsutus (hêr-sū′-tus) shaggy, rough with hair or prickles.
hirtus (hêr′-tus) rough, uneven, hairy.
Hirundo (hir-un′-dō)
hispanicus (his-pā′-ni-kus) Spanish.
hispidus (his′-pi-dus) rough, hairy, prickly.
Histiurus (his-ti-ū′-rus)
Histrionicus (his-tri-on′-ik-us)
Hodomys (hod′-ŏ-mis)
Hodotermitidae (hod-ŏ-têr-mit′-i-dē)
Hoffmannseggia* (hof-man-seg′-i-ạ)
Holacantha* (hol-ak-an′-thạ)
holandric (hol-an′-drik)
holarctic (hol-ârk′-tik, hōl-ârk′-tik)
holard (hō-lard′)
Holbrookia (hōl-brŏŏk′-i-ạ)
Holcochlaena* (hol-kok-lē′-nạ)
holcodont (hol′-kŏ-dont)
Holcus* (hol′-kus)
holoblastic (hol-ŏ-blas′-tik)
Holocera (hol-os′-er-ạ)
Holognatha (hol-og′-na-thạ)
Hololepta (hol-ŏ-lep′-tạ)

holomastigote (hol-ō-mas'-ti-gōt)
Holometabola (hol-ŏ-me-tab'-ŏ-lạ)
Holometopa (hol-ŏ-met-ōp'-ạ)
holophyllus (hol-of-il'-us) entire-leaved.
holophytic (hol-ŏ-fit'-ik)
holosericeus (hol-ŏ-sē-ris'-e-us) entirely silky.
Holosteum* (hol-os'-te-um)
Holothuroidea (hol-ŏ-thū-roy'-dê-ạ)
holotype (hol'-ot-īp)
holozoic (hol-ŏ-zō'-ik)
Homalium (hŏ-māl'-i-um)
Homalocladium* (hom-al-ŏ-kla'-di-um)
Homarus (hom'-ar-us)
homeosis (hom-ê-ō'-sis)
homoblastic (hom-ŏ-blast'-ik)
homocercal (hom-ŏ-sêr'-kal)
homodont (hom'-ŏ-dont, hō'-mō-dont)
homodromous (hom-od'-rŏ-mus)
homogenous (hom-oj'-ên-us)
homoiothermal (hom-oy-ō-thêr'-mal)
homologous (hom-ol'-ŏ-gus)
homologue (hom'-ŏ-log)
homomallus (hom-om-al'-us)
homospory (hom-os'-pôr-i)
homozygous (hom-ŏ-zī'-gus)
homunculus (hom-un'-ku-lus) a little man.
Hoplisodes (hop-lis-ōd'-ēz)
Hoplonemertea (hop-lŏ-nē-mêr'-te-ạ)
hordeiformis (hôr-de-i-fôr'-mis) with form like
   barley.
Hordeum* (hôr'-de-um)
horminum (hôr-mī'-num) a kind of sage.

Hormiphora (hôr-mi′-fôr-ạ)
hormones (hôr′-mōnz)
horridus (hor′-id-us) standing on end, shaggy, bristly; also, wild, horrid.
hortensis (hôr-ten′-sis) cultivated in gardens.
hospitable (hos′-pit-ab'l)
Hosta* (hos′-tạ)
Hovenia* (hō-ven′-i-ạ, hō-vēn′-i-ạ)
huia (hōō′-yạ)
huisache (wĕ-sä′-chȧ)
humifusus (hum-if-ū′-sus) spread out on the ground.
humilis (hum′-i-lis) low, small.
Humulus* (hū′-mu-lus)
Hyacinthus* (hī-a-sin′-thus)
Hyalonema (hī-à-lŏ-nē′-mạ)
hyaloplasm (hī′-al-ŏ-plazm)
hybrid (hī′-brid)
hybridization (hī-brid-i-zā′-shun)
hybridus (hī′-bri-dus) hybrid.
hydatid (hī′-dȧ-tid)
hydatiform (hī-dat′-i-fôrm)
Hydranassa (hī-dran-as′-ạ)
Hydrangea* (hī-dran′-jê-ạ)
Hydrobates (hī-drob′-ạ-tēz)
Hydrobius (hī-drob′-i-us)
Hydrocharis* (hī-drok′-à-ris)
Hydrochoerus (hī-drŏ-kē′-rus)
Hydrocleis* (hīd′-rŏk-līs)
Hydrocotyle* (hī-drŏ-ko′-ti-lē)
Hydrolea* (hī-drŏ-lē′-ạ, hid-ro′-lê-ạ)
Hydromantes (hī-drŏ-man′-tēz)

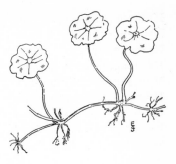

Hydrocotyle <Gr. *hydor*,
water+*kotylē*, a cavity or cup.
Pronounced: hī-drŏ-kot'-i-lē;
also, hid-rok-ot'-il-ē.

Hydrometra (hī-drom-et'-rạ)
Hydrophilidae (hī-drŏ-fil'-i-dē)
Hydrophilus (hī-drof'-il-us)
hydrophobia (hī-drŏ-fō'-bi-ạ)
Hydrophyllum* (hī-drŏ-fil'-um)
Hydropotes (hī-drop'-ŏ-tēz)
Hydroprogne (hī-drop-rog'-nē)
Hydroptila (hī-drop'-ti-lạ)
Hydroptilidae (hī-drop-til'-i-dē)
Hydroscapha (hī-dros-kā'-fạ)
Hyemoschus (hī-e-mos'-kus)
hygiene (hī'-ji-ēn, hī'-jēn)
Hyla (hī'-lạ)
hylaeus (hī-lē'-us) one of the centaurs.
Hylaplesia (hī-là-plē'-si-ạ)
Hylephila (hī-lef'-il-ạ)
Hylobates (hī-lob'-à-tēz)
Hylocharis (hī-lok'-à-ris)
Hylocichla (hī-lŏ-sik'-lạ)
Hylotoma (hī-lôt'-ŏm-ạ)
hymenium (hī-mēn'-i-um) a membrane.
Hymenocallis* (hī-men-ŏ-kal'-is)

Hymenolepis (hī-men-ol'-e-pis)
Hymenopappus* (hī-men-ŏ-pap'-us)
Hymenophyllum* (hī-men-ŏ-fil'-um)
Hymenoptera (hī-men-op'-têr-ạ)
Hymenorus (hī-men'-ôr-us)
Hymenoxys* (hī-men-oks'-is)
Hyoscyamus* (hi-os-si'-am-us, hī-ŏ-sī'-à-mus)
Hypatus (hip'-a-tus)
hypaxial (hip-aks'-i-al)
Hypena (hī-pē'-nạ)
Hypericum* (hip-er'-ik-um)
Hyperoödon (hī-pêr-ō'-ŏ-don)
Hyperotreta (hī-pêr-ŏ-trē'-tạ)
hyperpnoea (hī-pêr-nē'-ạ)
hypertrophy (hī-pêr'-tro-fi)
Hyphaene* (hī-fē'-nē, hif-ē'-nē)
Hyphantria (hī-fan'-tri-ạ)
Hyphanturgus (hī-fan-tūr'-gus)
Hypilate* (hip-i-lā'-tē)
hypnody (hip'-nŏ-di)
hypnoides (hip-no-ī'-dēz) resembling *Hypnum*, the
    feather-moss.
Hypnum* (hip'-num)
Hypocera (hī-pos'-er-ạ)
Hypochera (hī-pok'-êr-ạ)
Hypochilus (hī-pŏ-kīl'-us)
Hypocolius (hī-pŏ-kō'-li-us)
hypocotyl (hī'-pok-ot-il)
hypogaeous (hī-pŏj-ē'-us)
hypogaeus (hī-pŏj-ē'-us) underground.
hypogastric (hī-pŏ-gas'-trik, hip-ŏ-gas'-trik)
Hypohippus (hī-pŏ-hip'-us)

Hypohomus (hī-pō'-ho-mus)
Hyponomeutidae (hī-pŏ-nom-ū'-ti-dē)
hypophaeus (hī-pŏ-fē'-us) dusky below.
hypophloeodal (hī-pŏ-flē-ō'-dal)
Hypoprepia (hī-pŏ-pre'-pi-ạ)
Hyporhagus (hī-pŏ-rā'-gus)
Hypositta (hī-pos-it'-ạ)
hypothalamus (hī-pŏ-thal'-à-mus)
Hypotricha (hī-pot'-ri-kạ)
Hypoxis* (hī-poks'-is)
Hypsiglena (hip-si-glēn'-ạ)
Hypsilophodon (hip-si-lof'-ŏ-don)
Hypsiprymnodon (hip-si-prim'-nŏ-don)
Hyracotherium (hī-ra-kŏ-thē'-ri-um)
hyssopifolium (his-op-i-fol'-i-um, his-op-i-fō'-li-um) with leaves like *Hyssopus.*
Hyssopus* (his-ō'-pus)
Hystrichopsyllidae (his-tri-kŏ-psi'-li-dē)
Hystrix* (his'-tri-ks)
hyther (hīth'-êr)

I

Iapygidae (ī-à-pij'-i-dē)
Ibalia (ib-ā'-li-ạ)
Iberidella* (ib-ē-rid-el'-ạ)
Iberis* (ī-bē'-ris)
Ibicella* (ī-bi-sel'-ạ)
Ibycter (ī-bik'-têr)
Icacina* (ik-a-sī'-nạ)
Icaco (ik-ā'-kō)
Icerya (i-sēr'-i-ạ)
Ichneumia (ik-nū'-mi-ạ)

Ichneumonidae (ik'-nū-mon'-i-dē)
ichnite (ik'-nīt)
Ichnocarpus* (ik-nō-kâr'-pus)
Ichthyornis (ik-thi-ôr'-nis)
Ichthyosaur (ik'-thi-ŏ-sôr)
Icichthys (i-sik'-this)
icotype (ī'-kō-tīp)
Ictalurus (ik-tal-ū'-rus)
Icteria (ik-ter'-i-ạ, ik-tē'-ri-ạ)
Icterus (ik'-ter-us)
Icticyon (ik-tis'-i-ōn, ik-tis'-i-on)
Ictidomys (ik-tid'-ŏ-mis)
Ictiobus (ik-tī'-ŏ-bus)
Ictonyx (ik'-tŏ-niks)
id (id)
idant (id'-ant)
ideomotor (id-ē-ŏ-mō'-tôr)
Idesia* (īd-ē'-si-ạ)
idioandrosporous (id-i-ŏ-and-ros-pō'-rus)
idioblast (id'-i-ŏ-blast)
Idmonea (id-mō'-nê-ạ)
Idolothripidae (ī-dol-ŏ-thrip'-i-dē)
idoneus (i-dō'-ne-us) fit, proper, suitable, sufficient.
Idotea (ī-dō'-tê-ạ, i-dŏ-tē'-ạ)
Idothea* (ī-doth'-e-ạ, ī-do-thē'-ạ)
Iduna (i-dū'-nạ)
ignavus (ig-nā'-vus) slow, slothful, inactive.
igneus (ig'-ne-us) fiery.
ignotus (ig-nō'-tus) unknown.
Ilex* (ī'-leks)
iliacal (i-lī'-ȧ-kal)

Illecebraceae* (il-es-ê-brā'-se-ē)

illecebrosus (il-es-eb-rō'-sus) full of allurement, attractive.

Illecebrum* (il-es'-ê-brum, il-es-eb'-rum)

Illicium* (il-ish'-i-um, il-is'-i-um)

Illigera* (il-ij'-e-rą)

Ilysanthes* (il-is-anth'-ēz)

imaginal (im-aj'-in-al)

imago (im-ā'-gō, pl. im-a'-ji-nēz)

Imantophylum* (im-ant-of'-il-um)

imantus (im-ant'-us) a strap or throng.

imberbis (im-bêr'-bis) beardless.

imbricatus (im-brik-ā'-tus) overlapping, as if covered with tiles.

immutabilis (im-ū-tā'-bi-lis) changed, altered.

impar (im'-pâr) uneven, unequal, unlike, odd.

imparilis (im-par'-il-is) unlike, unequal.

imparipinnate (im-pâr-i-pin'-āt)

imparis (im'-par-is) unequal, uneven, odd; also, inferior.

Impatiens* (im-pā'-shi-enz)

impavidus (im-pav'-id-us) fearless.

imperialis (im-per-i-ā'-lis) kingly.

impiger (im'-pi-jêr) active, quick.

implexus (im-pleks'-us) plaited, interwoven.

impolitus (im-pol-ī'-tus) rough, not polished.

impotent (im'-pŏ-tent)

Inachidae (in-ak'-i-dē)

inaquosus (in-ak-wō'-sus) lacking water.

incanus (in-kān'-us) hoary.

incarnatus (in-kâr-nā'-tus)

incessus (in-ses'-us) a going, walking.

incisor (in-sī′-zêr, in-sī′-sêr, in-sī′-sôr)

incisum (in-sī′-sum) cut into.

incitatus (in-sit-ā′-tus) rapid, quick.

inclarus (in-klā′-rus) obscure.

incolatus (in-kol-ā′-tus)   dwelling in a place.

incongruent ((in-kong′-grū-ent)

increpitus (in-kre′-pi-tus) making a noise, rattling,
   rebuking.

incubaceus (in-kub-ā′-se-us) lying close to the
   ground.

incurvus (in-kêr′-vus) bent, curved.

indecoris (in-dek′-ôr-is) unbecoming, inglorious.

index (in′-deks, pl. in′-di-sēz)

indicus (in′-dī-kus) of India or the East Indies.

indigen (in′-di-jen)

Indigofera* (in-di-gof′-êr-ạ)

indivisus (in-dī-vī′-sus) undivided.

indumentum (in-dū-men′-tum)

indusium (in-dū′-zi-um, in-dū-si-um; pl. in-dus′-
   i-ạ)

inebriate (in-ē′-bri-āt)

inermis (in-êr′-mis) unarmed, defenseless.

infantile (in′-fan-tīl, in′-fan-til)

infectivus (in-fek-tī′-vus) pertaining to dyes.

infestus (in-fes′-tus) unsafe, infested, troublesome.

inflatus (in-flā′tus) blown up, inflated.

infuscatus (in-fus-kā′-tus) obscure.

ingenitus (in-jen′-i-tus) innate, instilled by birth.

ingens (in′-jenz) huge, enormous; also, remarkable.

ingluvies (in-gloo′-vi-ēz)

inhaerens (in-hē′-renz) connected, hung to.

inhalant (in-hā′-lant)

inherent (in-hē'-rent)
inion (in'-i-on)
innatus (in-ā'-tus) unborn.
innoxius (in-oks'-i-us) harmless, blameless, innocent; uninjured.
Inocarpus* (ī-nŏ-kâr'-pus, ī-nok-âr'-pus)
Inocellia (in-os-el'-i-ạ)
Inocelliidae (in-o-sel-ī'-i-dē)
Inodes* (in-ō'-dēz)
inodorus (in-od-ō'-rus) without smell.
inopinatus (in-op-in-ā'-tus) unexpected.
inopinus (in-op-īn'-us) unexpected.
inquiline (in'-kwi-lin)
insculptus (in-skulp'-tus) engraved, carved.
insignis (in-sig'-nis) remarkable, notable.
insolitus (in-sol'-i-tus) unusual, uncommon.
insonus (in'-son-us) without sound.
insperatus (in-spēr-ā'-tus) unexpected.
inspiratory (in-spīr'-ȧ-tō-ri)
inspissate (in-spis'-āt)
instabilis (in-stab'-il-is) unsteady.
integument (in-te'-gŭ-ment)
intercalary (in-têr-kal'-a-ri)
interdictus (in-têr-dik'-tus) prohibited.
intermedius (in-têr-med'-i-us) intermediate.
interpres (in-têr'-pres) an interpreter, an explainer, a go-between.
interruptus (in-têr-up'-tus) broken, parted, interrupted.
intestine (in-tes'-tin)
intine (in'-tin, in'-tīn)
intortus (in-tôr'-tus) twisted.

intutus (in-tū'-tus) defenseless, dangerous.

intybus (in'-ti-bus) chicory.

Inula* (in'-ū-lạ)

inundatus (in-und-ā'-tus) over-flowed.

invictus (in-vik'-tus) unconquered, invincible.

invisus (in-vī'-sus) unseen, unknown.

involucre (in-vo-lū'-kêr)

involutus (in-vol-ū'-tus) intricate, obscure, involved.

Io* (i'-ō)

Iochroma* (i-ok-rō'-mạ)

Ionidium* (i-on-id'-i-um)

ionoglossus (i-on-ȯ-glō'-sus, i-on-ȯ-glos'-us) violet-tongued.

Ionopsis* (i-on-op'-sis)

ionoptera (i-on-op'-têr-ạ)

Ionornis (i-on-ôr'-nis)

Iphiclides (if-ik-lī'-dēz)

Iphisa (if'-is-ạ)

Ipomoea* (ī-pȯ-mē'-ạ, ī-pom-ē'-ạ)

Ipsea* (ip'-se-ạ)

Irena (ī-rē'-nạ)

Iresine* (i-res-ī'-nē, i-rês-ī'-nē)

Iridoprocne (ir-i-dȯ-prok'-nē)

irrectus (ir-ekt'-us) not straight.

irriguus (ir-i'-gū-us) wet, swampy, full of water.

irrorate (ir'-ȯ-rāt)

irroratus (ir-ō-rā'-tus) moistened.

Isandra* (is-an'-drạ)

Isaria* (is-ā'-ri-ạ)

Isatis* (ī'-sȧ-tis, ī-sā'-tis)

Ischarum* (isk'-ar-um)

ischiocerite (is-ki-os′-e-rīt)
ischium (is′-ki-um)
Ischnocera (isk-nos′-er-ạ)
Ischnochiton (isk-nȯ-kī′-ton)
Ischnopsyllidae (isk-nȯ-psil′-i-dē)
Ismelia* (is-mē′-li-ạ)
Ismene* (is-mē′-nē)
Isnardia* (is-nârr′-di-ạ)
Isocardia (ī-sȯ-kâr′-di-ạ)
Isocoma* (ī-sȯ-kō′-mạ)
Isocybus (ī-sos-īb′-us)
isocytic (ī-sȯ-sit′-ik)
Isoetes* (ī-sȯ′-ê-tēz, īs-o′-ê-tēz)
isolecithal (ī-sȯ-les′-ith-al)
Isolepis* (ī-sȯl′-ep-is, ī-sol′-ep-is)
Isolobodon (ī-sō-lob′-ȯ-don)
Isoloma* (ī-sȯ-lō′-mạ)
Isomeris* (is-om′-er-is)
Isoodon (ī-sȯ′-ȯ-don)
isophyllus (ī-sof-il′-us) equal-leaved.

Isomeris <Gr. *isos,* equal+*meris,* part.
Pronounced: is-om′-er-is, but Ī-som′-
er-is is acceptable.

Isopoda (ī-sop'-ŏd-ạ)
Isoptera (ī-sop'-têr-ạ)
Isopyrum* (ī-sṓ-pī'-rum)
Isora* (ī-sō'-rạ)
Isotria* (ī-sō'-tri-ạ)
Itea* (it'-ê-ạ, ī'-te-ạ)
iter (i'-ter, ī'-ter)
Ithaginis (ith-aj'-i-nis)
Ithomia (ith-ōm'-i-ạ)
Itonididae (it-on-id'-id-ē, it-ō̇-nid'-id-ē)
Iva* (ī'-vạ)
Ixiolirion* (iks-i-ol-ī'-ri-on)
Ixobrychus (iks-ŏb'-rik-us)
ixocarpus (iks-ō̇-kâr'-pus) sticky-fruited.
Ixodia* (iks-ō'-di-ạ)
Ixora* (iks-ō'-rạ)
Ixoreus (ik-sôr'-e-us)

J

Jacana (hä'-kä-nä)
Jacaranda* (jak-a-ran'-dạ)
jackal (jak'-al)
Jacquemontia* (jak-kwem-ōn'-ti-ạ)
jaeger (yā'-ger)
jaguar (jag'-wär)
jalapa (jal'-ap-ạ)
Jaltomata* (jal-tom'-at-ạ)
Jalysus (jā'-lis-us)
Jambosa* (jam-bō'-sạ)
Jamesia* (jām'-zi-ạ)
Janipha* (jan-ī'-fạ)
Janthina (jan'-thi-nạ)

Jamesia. Named in honor of Dr. Edwin James, American botanist who discovered the plant. Pronounced: jăm′-si-ạ, not jā-mē′-si-ạ.

Janusia* (ja-nu′-si-ạ)
Jasione* (jas-i-ō′-nē)
Jasminum* (jas′-min-um)
Jassidae (jas′-i-dē)
Jatamansi* (jat-am-an′-si)
Jatropha* (jat′-rô-fạ, jā′-trof-ạ)
jecoral (jek′-ô-ral)
Jerboa (jêr-bō′-ạ, jer′-bō-ạ)
Jongheana* (jon-gē-a′-nạ)
Jubaea* (jū-bē′-ạ, jub-ē′-ạ)
jubatus (jub-ā′-tus) crested, having a mane.
Juglans* (jū′-glanz, jūg′-landz)
jugular (jū′-gū-lâr)
jugum (jū′-gum)
jujuba (jū′-jub-ạ, jū′-jūb-ạ)
jujubinus (jū-jub′-i-nus) jujub-like.
Julus (jū′-lus)
junceus (jun′-se-us) made of rushes, rush-like.
Juncus* (jun′-kus)
Juniperus* (jŭ-nip′-er-us)

Jurassic (jŭ-ras′-ik)
Justicia* (just-is′-i-ạ)
Jussiaea* (jus-si-ē′-ạ)
juvenal (jū′-ven-al)
juvencus (juv-enk′-us) young.
juvenile (jū′-ven-īl)

K

Kallstroemia* (kal-strē′-mi-ạ)
Kalopanax* (ka-lop′-an-aks)
Kalosanthes* (kal-os-anth′-ēz)
Kalotermitidae (ka-lô-têr-mit′-i-dē)
Kapala (kap′-al-ạ)
karyoplasm (kâr′-i-ô-plazm)
Kastnia (kast′-ni-ạ)
kenenchyma (ken-eng′-kim-ạ)
Kermes (kêr′-mēz)
Kielmeyera (kēl-mī-′êr-ạ)
kinesiatrics (kin-ē-si-at′-riks)
kinesis (kin-ē′-sis)
kinesodic (kin-ēs-od′-ik)
kinetochore (kin-et′-ô-kôr, kin-ē′-tô-kôr)
Kinixys (kin-iks′-iz)
kinkajou (king′-ka-jū)
Kinosternon (kin-ô-stêr′-non)
knephoplankton (nef-ô-plangk′-ton)
Kniphofia* (nif-of′-i-ạ)
Kobus (kō′-bus)
Kochia* (kuk′-i-a, kō′-ki-ạ)
Koeberlinia* (kēb-êr-lin′-i-ạ)
Koelreuteria* (kēl-roo-te′-ri-ạ)
Kogia (kō′-ji-ạ)

Kolkwitzia* (kolk-wit'-zi-ą)
Krameria* (krā-mer'-i-ą)
krummholz (krum'-hōlz)
Kyrthanthus* (kir-tan'-thus)

L

labellum (lab-el'-um)
labial (lā'-bi-al)
Labiatae (lā-bi-ā'-tē, lab-i-ā'-tē)
labiatus (lab-i-ā'-tus)
Labichea* (lab-ī'-ke-ą)
labidophorus (lab-i-dof'-ō-rus)
Labidura (lab-i-dūr'-ą)
Labiduridae (lab-i-dū'-ri-dē)
labidus (lā'-bid-us) slippery.
Labiidae (làb-ī'-i-dē)
labilis (lā'-bil-is) slipping, transient.
labium (lab'-i-um, lā'-bi-um)
Lablavia* (lab-lā'-vi-ą)
laboratory (lab-ôr-at-ō'-ri, lab'-ôr-ā-tō-ri)
Labrax (lā'-braks)
labrosus (lab-rō'-sus) thick-lipped.
labrum (lab'-rum, lā'-brum)
Laburnum* (lab-ur'-num)
Lacaena* (las-ē'-ną)
Laccobius (lak-ōb'-i-us)
Lacepedea* (las-ē-pē'-dê-ą)
lacerans (las'-er-anz) mutilating, torturing.
Lacerta (làs-êr'-tą)
Lacertilia (las-êr-til'-i-ą)
Lachenalia* (lak-en-al'-i-ą)
Lachesis (lak'-ê-sis)

Lachnaea (lak-nē'-ạ)
Lachnanthes* (lak-nanth'-ēz)
Lachnosterna (lak-nŏ-stêr'-nạ)
Lachnostoma (lak-nos'-tom-ạ)
laciniatus (las-in-i-ā'-tus) with jagged edges.
Lacistema* (lak-is-tē'-mạ)
Lacosoma (lak-os-ōm'-ạ)
lacrimal (lak'-ri-mal)
Lactuca* (lak-tū'-kạ)
lacuna (la-kū'-nạ, pl. la-kū'-nē)
lacunar (la-kū'-nâr)
lacunose (la-kū'-nōs)
lacustris (lak-us'-tris) associated with lakes or ponds.
Ladanum* (lā'-dan-um, lad'-à-num)
Laemobothriidae (lē-mŏ-both-rī'-i-dē)
laenatus (lē-nā'-tus) cloaked.
laetivirens (lē-tiv'-ir-enz) with bright-green foliage.
laetus (lē'-tus) cheerful, gay, pleasing, beautiful.
laevicaulis (lē-vik-ô'-lis)
laevigatus (lē-vi-gā'-tus) made smooth, smooth.
laevis (lē'-vis) smooth, slippery, soft.
lagena (laj-ē'-nạ) a flask
Lagenaria* (laj-ê-nā'-ri-ạ)
lagenarius (laj-ê-nā'-ri-us) of or pertaining to a bottle or flask.
lageniform (laj-ē'-ni-fôrm)
Lagenorhynchus (laj-ên-ŏ-ring'-kus)
Lagerstroemia* (lā-gêr-strē'-mi-ạ)
Lagidium (là-ji'-di-um)
Lagoa (lā-gō'-ạ)
Lagomorpha (la-gŏ-môr'-fạ, lå-gŏ-môr'-fạ)

Lagomys (lag'-ȯ-mis, lȧ-gō'-mis)
lagopinus (lag-ȯ-pī'-nus) like a hare's paw.
Lagopus* (lag'-ȯ-pus, la-gō'-pus)
Lagostomus (lag-os'-tȯ-mus)
Lagothrix (lag'-ȯ-thriks)
Lagunaria* (lag-ū-nā'-ri-ạ)
Lagurus (lag-ūr'-us)
lamella (lam-el'-ạ)
lamellar (la-mel'-âr, lam'-e-lâr)
Laminaria* (lam-in-ār'-i-ạ)
Lamium* (lā'-mi-um)
Lampranthus* (lam-pran'-thus)
lamprocarpus (lam-prȯ-kâr'-pus) shining fruit.
Lamprocolius (lam-prȯ-kō'-li-us)
Lamprogale (lam-prog'-a-lē)
Lampropeltis (lam-prȯ-pelt'-is)
Lamprotes (lam'-prȯ-tēz)
Lampyridae (lam-pir'-i-dē)
Lampyris (lam'-pir-is)
lanate (lā'-nāt)
lanatus (lā-nā'-tus) wooly, furnished with wool.
lanceolatus (lan-se-ol-ā'-tus) armed with a small point or lance.
Laniidae (lān-ī'-i-dē)
Lanius (lān'-i-us)
Lanivireo (lān-i-vir'-ė-ō)
lanose (lā'-nos)
Lantana (lan-tā'-nạ)
lanuginosus (lan-ū-jin-ō'-sus) woolly, full of down.
lanuginous (lan-ū'-jin-us)
lanugo (lan-ū'-gō, lā-nū'-gō)

lapathifolius (lap-ath-i-fol′-i-us, lap-ath-i-fō′-li-us)
  sorrel-leaved.
Laphria (laf′-ri-ạ)
Laphygma (lā-fig′-mạ)
lapideus (lap-id′-e-us) of stone, stony, a stone.
lapillus (lap-il′-us) a pebble.
Laplacea* (lap-lā′-se-ạ)
lappaceous (la-pā′-shus)
Lapponum* (lap-ō′-num)
Lapula* (lap′-ul-ạ)
largus (lar′-gus) abundant, large.
laricinus (lar-is′-in-us) larch-like.
Lariidae (lâr-ī′-i-dē)
Larix* (lar′-iks, lā′-riks)
Larrea* (lar′-e-ạ)
Larus (lā′-rus)
lascivus (las-i′-vus) playful, frisky.
Laserpitium* (las-êr-pish′-i-um, las-er-pit′-i-um)
Lasiandra* (las-i-an′-drạ)
Lasiocampidae (las-i-ȯ-kam′-pi-dē, lā-si-ȯ-kam′-
  pi-dē)
lasiolaenus (las-i-ȯ-lē′-nus, lā-si-ȯ-lē′-nus) shaggy
  cloak.
Lasionycteris (las-i-ȯ-nik′-têr-is, lā-si-ȯ-nik′-ter-is)
lasiophyllus (las-i-ȯ-fil′-us, lā-si-ȯ-fil′-us) shaggy-
  leaved.
Lasiopyga (las-i-ȯ-pi′-jạ, lā-si-ȯ-pi′-jạ)
Lasiosphaeria* (las-i-ȯ-sfē′-ri-ạ, lā-si-ȯ-sfē′-ri-ạ)
Lasiurus (las-i-ū′-rus, lā-si-ū′-rus)
Lasius (las′-i-us, lā′-si-us)
Latania* (lat-ā′-ni-ạ)
Latax (lā′-taks)

latebra (lat-eb′-rạ) a hiding place.

latebrosus (lat-eb-rō′-sus) obscure, secret, full of
  lurking places.

latex (lā′-teks)

Lathraea* (lath-rē′-ạ)

Lathyrus* (lath′-ir-us)

latidens (lā′-ti-denz) broad-toothed.

Lathyrus <*lathyros*, an old Greek name
for the pea. Pronounced: lath′-ir-us,
not lath-ī′-rus.

latifolius   (lā-ti-fol′-i-us,   lā-ti-fō′-li-us)   broad-
  leaved.

latipes (lā′-ti-pēz) broad-footed.

latiusculus (lāt-i-us′-ku-lus) somewhat broad.

latrans (lā′-tranz) barking.

latus (lā′-tus) broad.

latus (lā′-tus) carried, borne.

latus (*n.* la′-tus) the side, a lateral surface.

Lavatera* (la-vā-tē′-rạ)

Lavinia (la-vin′-i-ạ)

laxus (laks′-us) wide, roomy, open.

Lebia (lē′-bi-ạ)

Lebistes (lê-bis′-tēz)
Lecanium (lē-kā′-ni-um)
Lecanora* (lek-an-ō′-rạ)
lechuguilla (lech-ō͞o-gē′-yạ, lech-ŏŏ-gēl′-yạ)
lecithin (les′-ith-in)
lecotropal (lek-ot′-rop-al)
lectotype (lek′-tô-tīp)
lectus (lek′-tus) brought together.
Lecythis* (lē′-sith-is, les′-i-this)
Leda (lē′-dạ)
ledifolius (lē-di-fol′-i-us, lē-di-fō′-li-us) with leaves
   like *Ledum*, the Laborador tea.
Ledum* (lē′-dum)
legatus (lē-gā′-tus) appointed, chosen.
legume (leg′-ūm, lê-gūm′)
Leimadophis (lī-mad′-ô-fis)
Leiolopisma (lī-ô-lop-iz′-mạ)
Leiophyllum* (lī-of-il′-um)
Leiothrix (lī′-ô-thriks)
Leiotulus* (lī-ot′-ul-us)
Leipoa (lī-pō′-ạ)
Leitneria* (līt-nē′-ri-ạ)
Lemaireocereus* (lê-mā-rē-ô-sē′-rē-us)
lemma (le′-mạ, pl. lem′-at-ạ)
Lemmus (lem′-us)
Lemna* (lem′-nạ)
Lemniscomys (lem-nis′-kô-mis)
Lemonias (lê-mō′-ni-as)
lendigerus (len-di′-jer-us) bearing kernels.
Lendyanus* (len-di-ā′-nus)
lentiginosus (len-ti-jin-ō′-sus) freckled, full of
   spots.

Lentiscus* (len-tis'-kus)
leoninus (le-o̅-nī'-nus) of or belonging to a lion,
  colored yellow.
Leontocebus (le-ont-o̅-se̅'-bus)
Leontodon* (le-on'-to̅-don)
Leonurus* (le-o̅-nū'-rus)
Lepachys* (lep-ak'-is)
Lepadomorpha (lep-ad-o̅-mȏr'-fa̧)
Lepas (le̅'-pas)
Lepidagathis* (lep-id-ag'-ath-is)
Lepidium* (lep-id'-i-um)
lepidocarpus (lep-id-o̅-kȃr'-pus) scaley fruited.
Lepidoptera (lep-i-dop'-tȇr-a̧)
Lepidosaphes (lep-id-os'-af-e̅z)
Lepismidae (lep-iz'-mi-de̅)
Lepomis (le̜-pō'-mis)
Leporidae (lep-ȏr'-i-de̅)
Leporillus (lep-ȏr-il'-us)
leporinus (lep-ȏr-ī'-nus) of a hare.
Leptinus (lep-tīn'-us)
Leptoceridae (lep-to̅-ser'-i-de̅)
Leptodeira (lep-to̅-dī'-ra̧)
Leptodira (lep-to̅-dī'-ra̧)
Leptodora (lep-tod'-o̅-ra̧)
Leptogyne* (lep-toj'-in-e̅)
Leptolophus (lep-tol'-o̅-fus)
Leptonycteris (lep-to̅-nik'-ter-is)
Leptophis (lept-ōf'-is)
Leptophlebiidae (lep-to̅-fleb'-ī-i-de̅)
Leptoptilus (lep-top'-ti-lus)
Leptospermum* (lep-tos-pȇr'-mum)
Leptotes* (lep'-tot-e̅z)

Leptothyrium* (lep-toth-ir′-i-um)
Leptotyphlops (lep-tŏ-tif′-lops)
Lepturus* (lep-tū′-rus)
Lepus (lē′-pus, lep′-us)
Leria (lē′-ri-ạ)
Lernaea (lêr-nē′-ạ)
Lespedesa* (les-pê-dē′-sạ)
Lestes (lē′-stēz)
Lestidae (les′-ti-dē)
Lethocerus (lēth-os′-er-us)
lethostigma (lēth-ŏ-stig′-mạ)
Leucaena* (lū-sē′-nạ)
leucania (lū-kā′-ni-ạ)
Leucauge (lŭ-kŏ′-jē)
Leucelene* (lū-sē-lē′-nē)
Leucocorryne* (lū-kok-ôr′-in-ē)
leucocyte (lū′-kŏ-sīt)
Leucoium* (lū-kō′-i-um)
Leucojum* (lū-kō′-jum)
leucon (lū′-kon)
leucophaearia (lŭ-kof-ē-ā′-ri-ạ)
leucophaeus (lŭ-kof-ē′-us) white+dusky or gray.
Leucopogon* (lŭ-kop-ō′-gōn)
leucopsis (lŭ-kop′-sis) white-faced.
Leucopsis (lŭ-kop′-sis)
leucorhoda (lŭ-kor′-od-ạ) white rose.
Leucosolenia (lŭ-kŏ-sŏ-lēn′-i-ạ)
Leucosticte (lŭ-kŏ-stik′-tē)
Leucothoe* (lŭ-koth′-ŏ-ē)
leucothorectis (lū-kŏ-thôr-ēk′-tis)
leucurus (lŭ-kū′-rus) white-tailed.
levator (lev-ā′-tôr)

lever (lev′-êr, lē′-vêr)
levigate (lē′-vi-gāt)
levigatus (lē-vi-gā′-tus) smooth.
levipes (lev′-i-pēz) light-footed.
levis (lev′-is) light, not heavy.
levis (lē′-vis) smooth.
Levisticum (lev-is′-tik-um)
levulose (lē′-vu-los, lev′-ŭ-lōs)
Leycesteria* (lā-ses-tē′-ri-ạ)
Liatris* (lī-ā′-tris)
libani (lib′-an-ī) of Lebanon
Libellulidae (li-be-lūl′-i-dē)
Libocedrus* (lī-bos-ēd′-rus, lib-os-ēd′-rus)
libriform (lib′-ri-fôrm)
Lichanura (lik-an-ūr′-ạ)
Lichenes* (lī-kē′-nēz)
Lichnis* (lik′-nis)
Lichonycteris (lik-ô-nik′-têr-is)
ligamentum (lig-à-ment′-um)
Ligularia* (lig-ul-ā′-ri-ạ)
Ligusticum* (lig-us′-ti-kum)
ligustrinus (lig-us-trī′-nus) of the kind of the privet.
Ligustrum* (lig-us′-trum)
Ligyda (lij′-i-dạ)
Ligyrus (lij′-i-rus)
Lilacis* (lī-lā′-sis)
Lilium* (lī′-li-um, lil′-i-um)
limaciform (lī-mā′-si-fôrm)
limatulus (lī-mā′tul-us) somewhat filed or polished.
Limax (lī′-maks)
limbatus (lim-bā′-tus) bordered, with a hem, or edge.

Limia* (lī′-mi-ạ)
Limicola (lī-mik′-ŏ-lạ)
Limicolae (lī-mik′-ŏ-lē)
Limnada (lim′-na-dạ)
Limnanthemum* (lim-nan′-the-mum)
Limnephilidae (lim-nê-fil′-i-dē)
limnetic (lim-net′-ik)
Limnetis (lim-nē′-tis)
limnobates (lim-nob′-ạ-tēz)
Limnobia (lim-nō′-bi-ạ)
Limnobium* (lim-nō′-bi-um)
Limnodea* (lim-nō′-dê-ạ)
Limnogale (lim-nog′-al-ē)
limnology (lim-nol′-ŏ-ji)
Limnothlipis (lim-noth′-li-pis)
Limonium* (lī-mō′-ni-um)
Limosa (lī-mō′-sạ)
Limosella* (lī-mos-el′-ạ)
limosus (līm-ō′-sus) slimy, full of mud.
Limulus (lim′-ul-us)
Linanthus* (lī-nan′-thus)
Linaria* (lī-nā′-ri-ạ)
linariaefolius (lī-nā-ri-ē-fol′-i-us, lī-nār-i-ē-fō′-li-us)
    with leaves like the toad-flax, *Linaria.*
lineatus (lī-ne-ā′-tus) made straight; also, striped.
lingulatus (lin-gu-lā′-tus) shaped like a tongue.
linicolus (lī-ni′-ko-lus) growing amongst flax.
linin (lī′-nin)
Linnaea* (lin-ē′-ạ)
linophyllus (lī-nof-il′-us) with leaves like flax
    (*Linum*).
Linum* (lī′-num)

Linanthus <Gr. *linon*, thread+*anthos*, flower. Pronounced: lī-nan'-thus, not lin-an'-thus.

Liodera (lī-od'-ê-rạ)
liolaenus (lī-ol-ē'-nus) smooth-cloaked.
Liolepis (lī-ol'-e-pis)
Liomys (lī'-ộ-mis)
Liopeltis (lī-ộ-pelt'-is)
Liotheidae (lī-oth-ē'-id-ē)
Liparis* (lip'-à-ris)
Liparia (lī-par'-i-ạ)
lipase (li'-pās)
Lipeurus (lip-ūr'-us)
lipoclastic (lip-ō-klas'-tik)
Liposcelis (lip-os'-sel-is)
lipotype (lī'-pộ-tīp)
lipoxenous (lī-pok'-sê-nus)
liquis (lī'-kwis) oblique.
Liriodendron* (lī-ri-od-en'-dron)
Liriope* (lī'-ri-op-ē)
Litargus (lit-âr'-gus)
Lithocolletes (lith-ok-ol-ēt'-ēz)
Lithocolletis* (lith-ok-ol-ē'-tis)

Lithodes (li-thō'-dēz)
Lithodidae (li-thod'-i-dē)
Lithospermum* (lith-os-pêr'-mum)
litigiosus (lī-tij-i-ō'-sus) quarrelsome.
litoral (lit'-ôr-al)
litoralis (lī-to-rā'-lis) belonging to the shore.
litoreus (līt-ôr'-e-us)
Litorina (lit-ô-rī'-nạ)
litorosus (lī-tôr-ō'-sus) of or on the shore.
litotes (lī-tō'-tēz)
Litsea* (lit-sē'-ạ)
Littonia* (lit-on'-i-ạ)
Littorella* (lit-ôr-el'-ạ)
lituatus (lit'-u-ā'-tus) forked.
litus (līt'-us) of the sea-shore.
lividus (lī'-vid-us) blue, lead-colored.
livius (lī'-vi-us) lead-colored.
lobatus (lob-ā'-tus) divided into or bearing lobes.
Lobelia* (lō-bē'-li-ạ)
Lobesia (lō-besh'-i-ạ)
Lobipes (lō'-bi-pēz)
Lobium (lō'-bi-um)
Lobivanellus (lō-bi-van-el'-us)
Lobosa (lô-bō'-sạ)
Lobostemon* (lō-bos-tē'-mon)
lobotes (lō-bō'-tēz) lobed.
lobular (lob'-û-lâr)
lobule (lob'-ūl)
lobus (lō'-bus)
locellate (lô-sel'-āt)
locellus (lô-sel'-us)
Lochia (lō'-ki-ạ)

Lochites (lŏ-kī'-tēz)
loculicidal (lok-ū-li-sīd'-al)
Locustidae (lō-kus'-ti-dē)
lodicule (lod'-i-kūl)
Lodoicea* (lod-ŏ-is'-e-ạ)
Loligo (lō-lī'-go)

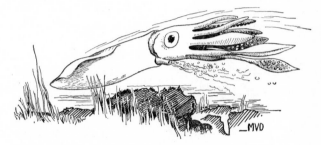

Loligo <L. *loligo*, a cuttlefish. Pronounced: lŏ-lī'-gō, not lo'-lī-gō.

Loligopsis (lōl-ī-gop'-sis)
Lolium* (lol'-i-um, lō'-li-um)
Lomaria* (lō-mā'-ri-ạ)
Lomariopsis* (lō-mā-ri-op'-sis)
Lomatium* (lō-mā'-shi-um, lō-mā'-ti-um)
Lomvia (lom'-vi-ạ)
Loncheres (long-kē'-rēz)
lonchochlamys (long-kok'-la-mis) with speared
    bracts.
Lonchophylla (long-kof-il'-ạ)
Lonchoptera (long-kop'-tĕr-ạ)
Lonchura (long-kū'-rạ)
longevity (lon-jev'-i-ti)
longiceps (lonj'-i-seps) long-headed.

longifolius (lon-ji-fol′-i-us, lon-ji-fō′-li-us) having long leaves.

longinquus (lon-jin′-kwu-us) long, extensive.

Longipennes (lon-jip-en′-ēz)

Lonicera (lon-is-ē′-rạ)

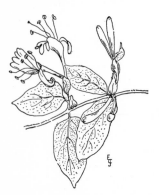

Lonicera. Named after Adam Lonicer (1528–1586), German botanist. Pronounced: lŏ-nis′-êr-ạ, also lon-is-ē′-rạ.

Lopezia* (lŏ-pēz′-i-ạ, lop-ē′-zi-ạ)

Lophanthus* (lŏ-fan′-thus, lof-an′-thus)

Lophiola* (lŏ-fi-ōl′-ạ, lof-i-ōl′-ạ)

Lophiomys (lŏ-fi′-ŏ-mis, lof-i′-ŏ-mis)

Lophocereus* (lŏ-fŏ-sē′-rê-us, lof-ŏ-sē′-rê-us)

Lophodytes (lŏ-fod′-i-tēz, lof-od-ī′-tēz)

Lopholatilus (lŏ-fŏ-lat′-i-lus, lof-ŏ-lat′-i-lus)

Lopholepis* (lŏ-fol′-ep-is, lof-ol′-ep-is)

Lopholithodes (lō-fŏ-lith′-ŏ-dēz, lof-ŏ-lith′-ŏ-dēz)

Lophopanopeus (lŏ-fŏ-pan-op′-e-us, lof-ŏ-pan-op′-e-us)

Lophophora* (lŏ-fof′-ôr-ạ, lof-of′-ôr-ạ)

Lophortyx (lŏ-fôr′-tiks, lof-ôr′-tiks)

Lophotes (lŏ-fō′-tēz, lof-ō′-tez)

Lophyrus* (lof-ī′-rus)
Lopimia* (lop-im′-i-ạ)
Lopus (lō′-pus)
Loranthus* (lō-ranth′-us)
lorica (lō-rī′-kạ, lōr′-i-kạ)
loriceus (lō-rī′-se-us) clothed in armor.
lotic (lō′-tik)
lotor (lō′-tôr) a washer.
Lottia (lot′-i-ạ)
Loxia (lok′-si-ạ)
Loxodonta (lok-sȯ-don′-tạ)
Loxotis* (loks-ō′-tis)
lubricus (lū′-brik-us) slippery.
Lucanidae (lū-kan′-i-dē)
Lucanus (lū-kān′-us)
lucens (lū′-senz) shining, conspicuous.
Lucernaria (lū-sêr-nā′-ri-ạ)
Lucidota (lū-si-dō′-tạ)
lucidus (lū′-si-dus) clear, full of light, bright.
Luciparens (lū-sip′-ar-enz)
lucius (lū′-si-us) a kind of fish.
luctuosus (luk-tu-ōs′-us) causing trouble, doleful.
luculentus (lū-ku-len′-tus) clear, bright, splendid.
Lucuma* (lū-kū′-mạ)
ludens (lū′-denz) sportive.
ludibundus (lū-di-bun′-dus) sportive, playful.
Luffa* (luf′-ạ)
lugubris (lū-gū′-bris) of or belonging to sorrow.
Luidia (lū-id′-i-ạ)
Luina* (lū′-in-ạ)
Lumbricus (lum-brī′-kus, lum′-bri-kus)
lumen (lū′-men, pl. lū′-mi-nạ)

Lunda (lun'-dạ)
lupine (lū'-pin)
lupinus (lup-ī'-nus, lū-pī'-nus)
lupulinus (lup-ū̇-lī'-nus) with habits or form of hops.
lupus (lup'-us) a wolf.
luridus (lū'-rid-us)
Luscinia (lū̇-sin'-i-ạ)
luscus (lus'-kus) one-eyed.
lusitanicus (lū-sit-ā'-nik-us) Portuguese, of Portugal.
lutarius (lu-tā'-ri-us) living on mud.
luteoalbus (lū-te-ō̇-al'-bus) yellowish-white.
luteolus (lū-te'-ol-us) yellowish.
luteus (lū'-te-us) yellow, golden-yellow, orange-yellow.
lutosus (lu-tō'-sus) full of mud, filthy, dirty.
Lutra (lū'-trạ)
Lutreola (lū-trē'-ol-ạ)
Luziola* (lū-zi'-ol-ạ)
Luzula* (lū'-zu-lạ)
Lycaena (lī-sē'-nạ)
Lycaenidae (lī-sen'-i-dē)
Lycaon (lis-ā'-on) an animal of the wolf kind.
Lychnis* (lik'-nis)
Lycioplesium* (lis-i-op-lē'-si-um)
Lycium* (lish'-i-um, lis'-i-um)
Lycogaster (lī-kog-as'-têr)
Lycoperdina (lī-kō̇-pêr-dī'-nạ)
Lycopersicon* (lī-kō̇-pêr'-si-kon)
Lycophyta (lī-kof'-it-ạ)
Lycopodiales* (lī-kō̇-pod-i-āl'-ēz, lī-kō̇-pō-di-āl'-ēz)

Lycium <Gr. *Lykion*, a name given to Rhamnus since it comes from Lycia. Pronounced: lis'-i-um, not li'-si-um.

Lycopodium* (lī-kop-od'-i-um, lī-kɔp-ō'-di-um)
Lycopsis* (lī-kop'-sis)
Lycopus (lī'-kŏ-pus)
Lycoris* (lī-kor'-is, lik-ō'-ris)
Lycornis (lī-kôr'-nis)
Lycosa (lī-kō'-sa, lik-ō'-sa)
Lycosidae (lī-kos'-id-e)
Lyctidae (lik'-ti-dē)
Lycurus* (lī-kūr'-us)
Lyda (lī'-da)
Lyencephala (lī-en-sef'-al-a)
Lygaeidae (lī-jē'-i-dē)
Lygeum* (lī-jē'-um)
Lygistum* (lij-is'-tum)
Lygodesmia* (lī-gŏ-des'-mi-a)
Lygodium* (lig-ō'-di-um, lī-gō'-di-um)
lygophil (lī'-gŏ-fil)
Lymantria (lī-man'-tri-a)
Lymantriidae (lī-man-trī'-i-dē)
Lymexylon (lī-meks'-il-on)
Lymnaea (lim-nē'-a)

Lyncea (lin-sē'-ạ, lin'-sē-ạ)
Lynx (links)
Lyonetiidae (lī-ŏ-net-ī'-i-dē)
Lyrocarpa* (lī-rŏ-kâr'-pạ, lir-ŏ-kâr'-pạ)
Lyroda (lī-rō'-dạ)
Lyrurus (lī-rū'-rus)
Lysichiton (lī-si-kī'-ton, lis-i-kī'-ton)
Lysiloma* (lī-si-lō'-mạ, lis-i-lō'-mạ)
Lysimachia* (lī-si-mā'-ki-ạ, lis-i-mā'-ki-ạ)
lysin (lī'-sin)
Lysiphlebus (lī-sif-lē'-bus)
Lyssianassidae (lis-i-a-nas'-i-dē)
Lythrum* (lith'-rum, lī'-thrum)
Lyurus (lī-ū'-rus)

M

Maba* (mā'-bạ)
Macaca (mak-ā'-kạ)
Macacus (mak-ā'-kus)
macaque (mā-käk')
macellarius (mas-el-ā'-ri-us) of or belonging to a
  meat-seller.
macer (ma'-ser) meager, lean.
Machaerocereus* (mak-ē-rŏ-sē'-re-us)
Machairodus (mak-ī'-rŏd-us)
Machilidae (mak-il'-i-de)
Macodes* (mak-ō'-dēz)
Macoma (mak-ō'-mạ)
Macradenia* (mak-rad-ē'-ni-ạ)
macradenous (mak-rad-ēn'-us) large-glanded.
Macranoplon* (mak-ran-op'-lon)
Macrochelys (mak-rok'-e-lis)

Macrochires (mak-rȯ-kī'-rēz)
Macrochloa* (mak-rok'-lo-ạ)
Macrocladus* (mac-rok'-lad-us)
Macratia (mak-rā'-ti-ạ)
Macrobasis (mak-rob'-as-is)
Macrogeomys (mak-rȯ-gē'-ȯ-mis)
Macronema* (mak-rȯ-nē'-mạ)
Macronyx (mak'-rȯ-niks)
Macrophya (mak-rof'-i-ạ)
Macroplethus* (mak-rop-lē'-thus)
Macropodidae (mak-rȯ-pod'-i-dē)
macropyrenic (mak-rȯ-pīr-ē'-nik)
macrorrhizus (mak-rȯ-rhī'-zus) with long or large roots.
Macroscelides (mak-ros-sel'-i-dēz, mak-rȯ-sel'-i-dēz)
Macrotus (mak-rō'-tus)
Macroxyela (mak-rȯ-zī'-el-ạ)
Macrozamia (mak-rȯ-zā'-mi-ạ)
maculatus (mak-ul-ā'-tus) spotted, speckled, dappled.
Madia* (mā'-di-ạ)
Madoqua (ma-dō'-kwạ)
Madreporaria (mad-rē-pȯr-ā'-ri-ạ, mad-rep-ȯr-ā'-ri-ạ)
madrepore (mad'-rē-pȯr)
madreporite (mad-rep'-ȯr-īt)
Magilus* (maj'-i-lus)
magnus (mag'-nus) large.
Mahonia* (mȧ-hō'-ni-ạ)
Maia (mā'-yạ)
maize (māz, mä-ēz')

Maianthemum* (mā-an′-the-mum, mā-yan′-thĕ-mum)

majalis (mā-jā′-lis) a gelded boar.

major (mā′-jôr) greater.

majus (māj′-us) great.

Malachium* (mal-ak′-i-um)

Malachius (mal-ak′-i-us)

Malaclemys=**Malaclemmys** (mal-a′-klem-is)

Malacomiza (mal-ak-om-īz′-ạ)

Malacostraca (mal-a-kos′-trȧ-kạ)

Malacothrix* (mal-a-kō′-thriks)

Malarcha* (mal-ârk′-ạ)

Malaxis* (mal-ak′-sis)

Malope* (mā′-lop-ē, mal′-ȯ-pē)

Malpighia* (mal-pig′-i-ạ)

maltose (mol′-tōs)

Malva* (mal′-vạ)

Malvastrum* (mal-vas′-trum)

Malvaviscus* (mal-vav-is′-kus)

Mammea* (mam-ē′-ạ)

Mammilaria* (mam′-i-lā′-ri-ạ)

Manaclus (man-ak′-lus)

Mandragora* (man-drag′-ôr-ạ)

manicatus (man-i-kā′-tus) furnished with long sleeves.

maniculatus (man-ik-ul-ā′-tus) with small hands.

Manolepis (man-ō′-lep-is)

Mantidae (man′-ti-dē)

mantis (man′-tis, pl. man′-tēz)

Mantispidae (man-tis′-pi-dē)

Marasmius* (mâr-as′-mi-us)

marcescent (mâr-ses′-ent)

marcianus (mâr-si-ā′-nus)
Mareca (mȧ-rē′-kạ)
margarine (mâr′-gâr-in)
margaritaceus (mâr-gâr-i-tā′-se-us) pearl-like.
Margarites (mâr-gâr-ī′-tēz)
Margarodidae (mâr-gȧ-rō′-di-dē)
marinus (mar-ī′-nus) of the sea, growing in the
waters of the sea.
maritimus (mar-it′-im-us) of or belonging to the
sea.
marmoratus (mâr-môr-ā′-tus) covered with marble.
Marmosa (mâr-mō′-sạ)
Marmota (mâr′-mõ-tạ)
Marrubium* (mar-ū′-bi-um)
marsupial (mâr-sū′-pi-al)
Martes (mâr′-tēz)
Masaridae (mas-a′-ri-dē)
Masaris (mas′-a-ris)
masculus (mas′-ku-lus) vigorous, manly, having
testicle-like tubers.
Masticophis (mas-tik′-ôf-is)
mastigium (mas-tij′-i-um)
Mastigophora (mas-ti-gof′-õ-rạ)
Mastotermitidae (mas-tõ-têr-mit′-i-dē)
Matricaria* (mat-ri-kā′-ri-ạ)
matrix (mā′-triks, pl. mā′-tri-sēz)
matronalis (mā-trōn-ā′-lis) of or belonging to a
married woman.
maturative (mat-ūr′-ȧ-tiv)
matutinal (mat-ū′-ti-nal)
Maurandya* (môr-an′-di-ạ)
maximus (maks′-im-us) largest, very large.

Mayaca* (mȧ-yak′-ạ, ma-yā′-kạ)
Mayetiola (mā-et-ī′-ol-ạ)
Maytenus* (mā′-ten-us, mā-tē′-nus)
Mazama (mä-zä′-mạ)
Mazus* (maz′-us)
Meandrina (mē-an-drī′-nạ)
means (me′-anz) going, passing; sometimes used in sense of quick-moving.
meatus (mē-āt′-us) a passage.
meconium (mē-kō′-ni-um)
Meconopsis* (mē-kōn-op′-sis)
Mecoptera (mê-kop′-têr-ạ)
Medeola* (mê-dē′-ol-ạ)
mediastinum (mē-di-as-tī′-num)
Medica* (mē′-dik-ạ)
Medicago* (mē-dik-ā′-go)
Medinilla* (mē-din-i′-lạ)
medius (me′-di-us) intermediate, in the middle.
medulla (med-ul′-ạ)
medullary (med′-ū-la-ri, mê-dul′-a-ri)
Medusa (me-dūs′-ạ)
Megaceryle (meg-a-sêr′-i-lē)
Megachile (meg-ȧ-kī′-lē)
Megachilidae (meg-ȧ-kil′-i-dē)
Megaderus (me-gad′-ê-rus)
Megadrili (meg-ȧ-drī′-lī)
Megalobatrachus (meg-ȧ-lô-bat′-rȧ-kus)
Megalodachne (meg-ȧ-lô-dak′-nē)
Megalodon (meg′-a-lô-don, meg-al′-ô-don)
Megalonyx (meg-a-lon′-iks)
Megalops (meg′-a-lops)
Megalopyge (meg-a-lop-ī′-jē)

Megalornis (meg-al-ôr′-nis)
Megaphyton (meg-af′-i-ton)
Megascops (meg′-ạ-skops)
Megaspilus (meg-as-pī′-lus)
Megathymus (meg-ath-īm′-us, meg-à-thī′-mus)
Megilla (mē-jil′-ạ)

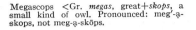

Megascops <Gr. *megas*, great+*skops*, a
small kind of owl. Pronounced: meg′-ạ-
skops, not meg-ạ-skōps.

—MVD

meiogenic (mī-ō̇-jen′-ik)
meiomery (mī-om′-êr-i)
meiosis (mī-ō-′sis)
meiotic (mī-ot′-ic)
Meiracylium* (mīr-ak-il′-i-um)
Melaleuca* (mel-al-ū′-kạ)
Melampodium* (mel-am-pō′-di-um)
Melampus (mel-am′-pus)
Melampyrum* (mel-am-pī′-rum)
Melanerpes (mel-an-êr′-pēz)
melanin (mel′-à-nin)
melanism (mel′-à-nizm)
melanistic (mel-an-is′-tik)
Melanitta (mel-an-it′-ạ)

Melanocarpum* (mel-an-ok-âr'-pum)
melanocorys (mel-an-ok'-ôr-is) black helmet.
melanophore (mel'-an-ô-fôr, mel-an'-ō-fôr)
Melanoplus (mel-an'-ô-plus)
Melanthium* (mel-an'-thi-um)
meleagridis (mel-ê-ā'-gri-dis) of the guinea-fowl.
Meleagris (mel-ê-ā'-gris)
Meleoma (mel-ê-ōm'-ạ)
Meles (mē'-lēz)
Melia* (mel'-i-ạ)
Meliantheae* (mel-i-anth'-ê-ē)
Melica* (mel'-i-kạ)
Melicope* (mel-ik'-op-ē)
melilot* (mel'-i-lot)
Melilotus* (mel'-i-lō'-tus)
Meliosma* (mel-i-os'-mạ)
Meliponidae (mel-i-pon'-i-dē)
Melissa* (mel-is'-ạ)
Melissodes (mel-is-ōd'-ēz)
Melittis* (mel-it'-is)
Melittobia (mel-it-ob'-i-ạ)
melleus (mel'-e-us) of honey, honey-sweet, de-
 lightful.
Mellivora (mel-iv'-ô-rạ)
Melocactus* (mel-ô-kak'-tus)
Melochia* (mel-ok'-i-ạ)
melodus (mel-ō'-dus) melodious.
Meloidae (mel-ō'-i-dē)
Melolonthidae (mel-ô-lon'-thi-dē)
Melophagus (mel-of'-ag-us)
Melospiza (mel-ô-spīz'-ạ)
Melothria* (mê-loth'-ri-ạ)

Membracidae (mem-brasʹ-i-dē)
Membranipora (mem-brā-nipʹ-ôr-ạ)
membranous (memʹ-brā̆-nus)
mendicus (men-dīʹ-kus) needy, beggarly.
menicatus (men-ik-āʹ-tuȝ) made into a crescent.
meningeal (men-inʹ-je-al)
meninges (men-inʹ-jēz)
Meniscotherium (men-is-kŏ-thēʹ-ri-um)
Menispermum* (men-i-spêrʹ-mum)
Menodora* (men-ŏ-dôrʹ-ạ)
Menoponidae (men-ŏ-ponʹ-i-dē)
Menotypla (men-ŏ-tipʹ-lạ)
Mentha* (menʹ-thạ)
Mentzelia* (ment-zēʹ-li-ạ)
Menura (men-ûrʹ-ạ)
Menyanthes* (men-i-anʹ-thēz)
Mephitis (mĕ-fīʹ-tis, mef-īʹ-tis)
Mercurialis* (mêr-kūr-i-āʹ-lis)
merens (merʹ-enz) deserving; also, guilty.
merganser (mer-ganʹ-ser)

Mephitis <L. *mephitis,* a pestilential exhalation. Pronounced:
mefʹ-it-is, not me-fitʹ-is.

mergens (mer'-jenz) dipped, sinking.
Mergus (mêr'-gus)
meridianus (mer-id-i-ā'-nus)
Meriones (mē-rī'-ȯ-nēz)
meroblastic (mer-ȯ-blas'-tik)
Meropidae (mē-rop'-i-dē)
Merops (mer'-ops, mē'-rops)
Mertensia* (mêr-ten'-si-ạ)
merulus (mer'-ul-us) a blackish bird.
Merychippus (mer-i-kip'-us)
mesaeum (mes-ē'-um)
mescal (mes-kal')
Mesembryanthemum* (mes-ēm-bri-anth'-em-um,
    mes-em-bri-anth'-em-um)
mesenchymal (mes-eng'-ki-mal)
mesenchyme (mes-eng'-kīm)
mesentery (mes'-en-ter-i)
mesepimeron (mes-e-pim'-ȇ-ron)
mesial (mē'-zi-al)
mesic (mes'-ik, mē'-sik) pertaining to the middle.
Mesites (mes-ī'-tēz)
mesoderm (mes'-ȯ-dêrm)
mesoglea (mes-ȯ-glē'-ạ)
mesomelas (mes-o'-me-las) halfway black.
Mesoplodon (mes-op'-lȯ-don)
Mesovelia (mes-ov-ēl'-i-ạ)
Mesozoic (mes-ȯ-zō'-ik)
Mespilus* (mes'-pil-us)
mesquite (mes-kēt'-ā̊, mes-kēt')
Mesua* (mē'-su-ạ, mes'-û-ạ)
Metachirops (met-ạ-kī'-rops)
metameric (met-a-mer'-ik)

metamerism (met-am′-er-izm)
Metandrocarpa (met-an-drŏ-kâr′-pą)
meteloides (met-el-o-ī′-dez) like metel, a kind of
plant.
Metepiera (met-e-pī′-rą)
Methoca (meth-ōk′-ą)
Metis (mē′-tis)
metoecious (met-ē′-shus)
Metopia (met-ōp′-i-ą)
Metopoceros (met-ŏ-pos′-er-os)
Metridium (mĕ-tri′-di-um)
Metrosideros* (mē-trō-si-dē′-ros, met-ros-id-ē′-
ros)
Metroxylon* (mĕ-troks′-il-on)
Meum* (mē′-um)
Mezira (mez-ī′-rą)
Miarchus (mī-âr′-kus)
micans (mik′-anz) glittering.
Micranthemum* (mī-kran′-the-mum)
micranthus (mī-kran′-thus)
Micrathene (mik-ra-thē′-nē)
Microcebus (mī-kro-sē′-bus)
Microdipodops (mī-krŏ-di′-pŏd-ops)
microdon (mī′-krŏd-on)
Microgadus (mī-krŏ-gā′-dus)
microglochin (mī-krŏ-glō′-kin) a small point.
microgyne (mī-kroj′-in-ē, mī′-krō-jīn)
Microligea (mī-krŏ-li′-je-ą)
Micromalthus (mī-krŏ-mal′-thus)
micromeris (mī-krom′-er-is) a small part.
Microrhagus (mī-krŏ-rag′-us)
micron (mī′-kron)

Micropalama (mī-krŏ-pal'-a-mạ)
Micropodidae (mī-krŏ-pod'-i-dē)
microscopist (mī-kros'-kŏ-pist)
Microseris* (mī-kros'-er-is)
Microsorex (mī-krŏ-sō'-reks)
Microstylis* (mī-kros'-til-is)
Microtus (mī-krōt'-us)
Micruroides (mīk-rū-ro-ī'-dēz)
Micrurus (mī-krū'-rus)
Midas (mī'-das)
Mididae (mid'-i-dē)
mignonette (min-yun-et')
Mikania* (mik-an'-i-ạ)
miliarius (mī-li-ā'-ri-us) of millet; also, containing
   a thousand.
militaris (mī-li-tā'-ris) war-like, like a soldier.
Milium* (mil'-i-um)
milleped (mil'-e-ped)
milpa (mil'-pạ)
milvus (mīl'-vus) a bird of prey, a kite.
Mimesidae (mī-mes'-id-ē)
mimetic (mi-met'-ik, mī-met'-ik)
Mimosa* (mī-mō'-sạ)
Mimulus* (mim'-ŭ-lus, mī'-mul-us)
Mimus (mī'-mus)
Mimusops* (mī'-mus-ops)
minax (mi'-naks) projecting.
minimus (min'-i-mus) very small, least, smallest.
Minois (min-ō'-is)
minor (mī'-nor) smaller
minus (mī'-nus) less, subtracting.
minute (*adj.* min-ūt')

Mimosa <L. *mimus*, an actor. Pronounced: mī-mō′-sạ, also sometimes pronounced mi-mō′-sạ, but this is not correct
but rather a pronunciation long used and
so accepted.

minutus (min-ū′-tus) small.

Miocene (mī′-o-sēn)

Miohippus (mī-ȯ-hip′-us)

miracidium (mī-rạ-sid′-i-um)

Mirafra (mir-af′-rạ)

Miridae (mīr-i-dē, mir′-i-dē)

mirificus (mī-ri′-fi-kus) wonderful, strange.

Mirounga (mir-oung′-gạ)

Mimulus <Late L. *mimulus* <L. *mimulus*, a diminutive <*mimos*, an actor. Pronounced: mī′-mul-us,
but mim′-ū-lus is almost always used.

mirus (mī'-rus) wonderful, extraordinary.
Miscophus (mis-kō'-fus)
miser (mis'-er) wretched.
mistletoe (mis'l'-tō)
Mitella* (mit-el'-ạ)
mitis (mī'-tis) mellow, ripe, soft, gentle.
mitiusculus (mī-ti-us'-kul-us) mild, very gentle.
mitochondria (mī-tộ-kon'-dri-ạ)
mitosis (mī-tō'-sis, mit-o'-sis)
mitral (mī'-tral)
mitralis (mī'-trā-lis) pertaining to a head-band or
    turban.
mitriform (mī'-tri-fôrm)
Mnemiopsis (nē-mi-ops'-is)
Mniotilta (nī-ộ-til'-tạ)
Mnium* (nī'-um)
Mobula (mob'-ử-lạ)
Modiola* (mộ-dī'-ộ-lạ, mod'-i-ol-ạ)
modiolus (mộ-dī'-ộ-lus)
Moeritherium (mēr-i-thē'-ri-um)
mola (mol'-ạ) a millstone.
Molamba (mol-am'-bạ)
Molanna (mol-an'-ạ)
Molannidae (mō-lan'-i-dē)
molecule (mol'-ê-kūl, mō'-lê-kūl)
Molge (mōl'-jē)
molitor (mol'-i-tôr) a grinder, a miller.
mollis (mol'-is) soft.
Mollugo* (mol-u'-gō)
Moloch (mō'-lok)
Molossus (mộ-los'-us)
Molothrus (mol'-ộ-thrus)

Momordica* (mom-ôr'-di-ką)
Momota (mŏ-mō'-ta)
Mompha (mom'-fą)
Monachus (mon'-ȧ-kus)
monad (mon'-ad, mō'-nad)
Monadina (mō-na-dī'-ną)
Monarda* (mon-âr'-dą)
Monarthrum (mon-âr'-thrum)
monax (mon'-aks) a monk.
Monedula (mon-ed'-ul-ą)
Moneses* (mon'-es-ēz, mŏ-nē'-sēz)
Monezia (mon-ēz'-ią)
Monilia* (mon-ī'-li-ą)
Moniliales (mon-ī-li-ā'-lēz)
moniliferus (mon-il-i'-fer-us) bearing a necklace or
   collar.
moniliform (mon-il'-i-fôrm)
monilis (mon-ī'-lis) of a necklace.
Monniera* (mon-i-ē'-rą)
monobasis (mon-ob'-as-is)
Monoclonius (mon-ŏ-klō'-ni-us)
monoecious (mŏ-nē'-shus, mon-ē'-shus)
Monogenea (mon-ŏ-jē'-nĕ-ą)
monogynus (mon-oj'-in-us) with single style.
monogyra (mon-ŏ-jī'-rą) single-whorled.
Monohammus (mon-ŏ-ham'-us)
monohybrid (mon-ŏ-hī'-brid)
Monolopia* (mon-ol-ō'-pi-ą)
Monopelis* (mon-op'-el-is)
Monophyllus (mon-ŏ-fil'-us)
Monotoma (mon-ot'-ŏ-mą)
Monotropa* (mon-ot'-rop-ą)

montanus (mon-tān'-us) belonging to a mountain,
  dwelling in mountains.
Montia* (mon'-ti-ạ)
monticolus (mon-ti'-kol-us) mountain-dweller.
Mopalia (mō-pāl'-i-ạ)
mopane (mô-pä'-nē)
Moraea* (môr-ē'-ạ)
moray (mō'-rā)
mordax (môr'-daks) given to biting, snarling.
Mordellidae (môr-del'-i-dē)
Mordellistena (môr-del-is'-ten-ạ)
Morina* (môr'-ī-nạ)
Moringa* (môr-in'-gạ)
Moris (mō'-ris)
Moronidae (mō-ron'-i-dē)
Moronobea* (môr-ô-nō'-bê-ạ)
Moropus (môr'-ô-pus)
Moroteuthis (mōr-ô-tū'-this)
Morphoidae (môr-fō'-i-dē)
morrhua (môr-ū'-ạ)
morula (môr'-ŭl-ạ)
Morus* (mō'-rus, môr'-us)
Mosasaurus (mō-sa-sô'-rus)
moschatus (mos-kā'-tus) having the odor of musk,
Moschus (mos'-kus)
motacilla (mō-ta-si'-lạ) the wagtail.
Motacillidae (mō-ta-sil'-i-dē)
mouflon (mo͞of'-lon)
mucronatus (mū-krō-nā'-tus) ending in a short
  point, pointed.
mucronis (mū-krōn'-is) of a sharp point or edge.
Muehlenbeckia* (mū-len-bek'-i-ạ)

Mugil (mū´-jil)
Mugilidae (mu-jil´-i-dē)
Muilla (mū-il´-ạ)
mulatto (mŭ-lat´-ō)
Mulgedium* (mul-jē´-di-um)
multicaulis (mul-ti-kô´-lis) many-stalked.
Mungos (mung´-os)
Munia (mū´-ni-ạ)
Muntiacus (mun-tī´-ak-us)
muralis (mū-rā´-lis) belonging to walls.
Murgantia (mûr-gan´-ti-ạ)
muricatus (mū-ri-kā´-tus) pointed.
Muridae (mū´-ri-dē)
Murinus (mū-rī´-nus)
murorum (mūr-ôr´-um) of walls.
murre (mêr)
Mus (mūs, mus)
Musa* (mū´-sạ, mū´-zạ)
Musaceae* (mū-sā´-sė-ē)
musang (mū-sang´)

Mus <L. *mūs*, mouse. Pronounced: mūs, but New Latin *mus* is considered
acceptable.

Muscardinus (mus-kâr-dī′-nus)
Muscari* (mus-kā′-ri)
muscariform (mus-kar′-i-form)
muscarius (mus-kā′-ri-us) belonging to flies.
Muscicapa (mus-ik′-ap-ạ)
Muscidae (mus′-i-de, mūs′-i-dē)
musciferus (mus-if′-er-us) bearing moss, moss-like.
muscipulus (mus-ip′-ul-us) fly-catching.
Muscivora (mus-iv′-ôr-ạ)
muscoides (mus-ko-ī′-dēz) like moss.
muscosus (mus-kō′-sus) moss-like, mossy.
musimon (mus′-i-mon)
Mustela (mus-tē′-lạ)
mustelinus (mus-tē-lī′nus) weasel-colored, of or belonging to a weasel.
muticus (mut′-i-kus) blunted, curtailed, lopped off.
Mutillidae (mū-til′-i-dē)
Myadestes (mī-ȧ-des′-tēz)
Mycetochares (mī-sēt-ok′-âr-ēz)
Mycetophagus (mī-sē-tof′-ȧ-gus)
Mycetophila (mī-sē-tof′-il-ạ)
Mycetophilidae (mī-sē-tŏ-fil′-i-dē)
Mycetozoa (mī-sē-tŏ-zō′-ạ)
Mycomyia (mī-kom-ī′-i-ạ)
Mycteria (mik-tē′-ri-ạ)
Mydaidae (mid-ā′-i-dē)
Mydaus (mid′-ȧ-us)
myelin (mī′-el-in)
myeloblast (mī-el′-ŏ-blast)
Mygale (mig′-ȧ-lē)
Myiarchus (mī-i-ârk′-us, mī-yârk′-us)

Myioborus (mī-i-ọ-bôr′-us, mī-yọ-bôr′-us)
Myiochanes (mī-i-ọ-kān′-ēz) mī-yọ-kān′ēz)
Myiodioctes (mī-i-ọ-di-ok′-tēz, mī-yọ-di-ok′-tēz)
Myiopsitta (mī-i-ọ-sit′-ạ)
Mymaridae (mī-mâr′-i-dē)
Myoporum* (mī-op′-ôr-um)
Myosorex (mī-os-ō′-reks)
Myosotidium* (mī-os-ọ-tid′-i-um)
Myosotis* (mī-os-ō′-tis)
Myosurus* (mī-os-ū′-rus)
Myotis (mī-ōt′-is)
Myriapoda (mir-i-ap′-ọ-dạ)
Myrica* (mir-ī′-kạ)
Myriodaria (mûr-i-ọ-dā′-ri-ạ)
Myriophyllum* (mir-i-ọ-fil′-um)
Myrmecobius (mûr-mē-kōb′-i-us)
Myrmecolacidae (mûr′-mē-kọ-las′-i-dē)
myrmecology (mûr-mē-kol′-o-ji)
Myrmecophaga (mûr-me-kof′-a-gạ)
Myrmeleontidae (mûr-mê-le-ont′-i-dē, mur-mê-lē-
  ont′-i-dē)
Myrmica (mûr-mīk′-ạ)
Myrrhis* (mir′-is)
myrsinites (mir-sin-ī′-tēz) myrtle-like.
Myrtillocactus* (mûr-til-ọ-kak′-tus)
Myrus (mī′-rus)
Mysis (mī′-sis)
mytilid (mī′-til-id)
Mytilus (mit′-il-us)
Myxine (miks-ī′-nē)
myxinoid (miks′-in-oyd)
Myxomycetes (miks-ọ-mī-sē′-tēz)

Myxomycophyta (miks-ō-mī-kof'-it-ạ)
Myzine (mī-zī'-nē)
Myzomela (mī-zom'-êl-ạ)
Myzostoma (mī-zos-tō'-mạ)
Myzus (mī'-zus)

N

Nabalus* (nab'-al-us)
Nabidae (nab'-i-dē
Nacerdes (nạ-sêr'-dêz)
nacre (nā'-kêr)
Naeogeus (nē-oj-ē'-us)
naevius (nē'-vi-us) spotted with moles, with blemishes.
naiad (nā'-yad, nī'-ad)
Naias* (nā'-yas)
Naja (nā'-ja)
Nama* (nā'-mạ)
Nannochoristidae (nan-ō-kô-ris'-ti-dē)
Nannus (nan'-us)
nanus (nā'-nus) a dwarf.
Napaea* (nā-pē'-ạ)
Napaeozapus (nā-pē-ô-zā'-pus)
napellus (nā-pel'-us) a little turnip.
Napus* (nā'-pus)
Narcine (nâr-sī'-nē)
Narcissus* (nâr-sis'-us)
Narcobatis (nâr-kob'-ạ-tis)
Narcomedusae (nâr-kô-mê-dūs'-ē)
Nardus* (nâr'-dus)
nares (nā'-rēz, sing. of nā'-ris)
Narthecium* (nâr-thē'shi-um, nâr-thē'-si-um)

Nasalis (nȧ-sāl'-is)
nascent (nas'-ent, nā'-sent)
nasicus (nā'-si-kus) nosed, with a nose.
Naso (nā'-sō)
Nasturtium* (nas-tûr'-shi-um)
nasus (nās'-us) nose.
nasutus (nā-sū'-tus) large-nosed.
natant (nā'-tant)
Nathodus (nath'-o-dus)
Natica (nat'-ik-ȧ)
Natrix (nā'-triks)
Naucinus (nô'-sin-us)
Nauclerus (nô-klē'-rus)
Naucoridae (nô-kôr'-i-dē)
naucrates (nô-krā'-tēz) a pilot.
navalis (nā-vā'-lis) belonging to ships.
Navarretia* (nav-âr-et'-i-ȧ)
navicular (nā-vik'-ụ-lâr)
neanderthalensis (nê-an-der-täl-en'-sis)
Nebalia (nê-bā'-li-ȧ)
necator (nek-ā'-tôr) a murderer.
Nectarophora (nek-târ-of'-ôr-ȧ)
Nectogale (nek-to'-gȧ-lē)
Nectria* (nēk'-tri-ȧ)
Necturus (nek-tū'-rus)
Neelidae (nê-el'-i-dē)
Neelus (nê-ēl'-us)
Negundo* (nē-gun'-dō)
Neides (nē-īd'-ēz)
Nelumbo* (nē-lum'-bō)
Nemacladus* (nê-mak'-la-dus)
Nemastylis* (nē-mas'-til-is)

Nemocladus <Gr. *nēma*, genit. *nēmatos*,
a thread+*klados*, a branch. Pronounced:
nē-mak'-la-dus, not nē-ma-klad'-us.

Nemathelminthes (nĕm-at-hel-min'-thēz)

Nematocera (nĕm-ȧt-os'-êr-ạ)

Nemocladus (nē-mak'-la-dus)

nematocyst (nĕm'-at-ȯ-sist)

Nematodirus (nĕm-at-ȯ-dī'-rus)

Nematomorpha (nĕm-at-ȯ-môrf'-ạ)

Nematus* (nē'-mat-us)

Nemertez (nē-mêr'-tēz)

Nemesia (nem-ē'-shi-ạ, ne-mē'-si-ạ)

Nemia* (nē'-mi-ạ)

Nemocera (nē-mos'-er-ạ)

Nemognatha (nē-mog'-nath-ạ, nem-og'-nath-ạ)

Nemopanthes* (nē-mop-an'-thēz)

Nemophila* (nē-mof'-il-a, nem-of'-il-ạ)

Nemopoda (nē-mop'-ȯd-ạ)

Nemopteridae (nē-mop-ter'-i-de, nem-op-ter'-i-dē)

nemoralis (nem-or-āl'-is) belonging to woods.

Nemorhaedus (nem-ȯ-rē'-dus)

nemorosus (nem-or-ō'-sus) full of foliage, bushy;
also, woody, shady.

Nemophila <Gr. *nemos*, a glade and *philos*, fond of. Pronounced: nem-of'-il-ạ.

nemorus (nem'-ôr-us) of woods, of groves.
Nemoseris* (nem-os'-er-is)
Nemospiza (nem-ộ-spī'-zạ)
Nemouridae (nem-ûr'-i-dē)
Neofelis (nē-of'-el-is)
Neofiber (nē-of'-i-bêr, nē-ộ-fī'-bêr)
Neogaea (nē-ộ-jē'-ạ)
Neognathae (nê-og'-na-thē)
Neomenia (nē-ộ-mēn'-i-ạ)
Neopasites (nê-ộ-pas-ī'-tēz)
Neophron (nē'-ộ-fron)
Neopieris* (nê-ộ-pī'-er-is)
Neosorex (nê-ộ-sō'-reks)
Neotinea* (nê-ot-in'-ê-ạ)
Neotoma (nē-ot'-ộ-mạ)
Neotremata (nē-ộ-trem'-a-tạ)
Nepa (nē'-pạ)
Nepenthes* (nē-pen'-thēz)
Nepeta* (nep'-et-ạ, nep'-ê-tạ)

Neotoma <Gr. *neo-*, new+*tomō*, to cut. Pronounced: nē̆-ot′-ŏ̆-mạ, not nē-ō-tō′mạ. The last *o* is not considered long, therefore it does not receive the accent.

Nephecoetes (nef-ē-sē′-tez)
Nephila (nef′-il-ạ)
nephridium (nef-rid′-i-um)
Nephrodium* (nef-rō′-di-um)
Nephrolepis* (nef-rol′-ep-is)
Nephropetalum* (nef-rŏ̆-pet′-al-um)
nephrostoma (nef-ro′-stŏ̆-mạ)
nephrostome (nef′-rŏ̆-stōm)
nepionic (nē-pi-on′-ik)
Nepticula (nep-tik′-ū-lạ)
Nepticulidae (nep-tik-ūl′-i-dē)
Nereis (nē′-rē̆-is)
Nereocystis (nē-rē̆-ŏ̆-sis′-tis)
Nerissa* (ner-is′-ạ)
Nerita (nē̆-rī′-tạ)
neritic (nē̆-rit′-ik)
neritinus (nē-rit′-in-us) like *Nerita*, a seamussel.
Nerium* (nē′-ri-um)

Nertera* (ner'-ter-ạ)
nesioticus (nē-si-ōt'-i-kus) belonging to an island.
Neslia* (nes'-li-ạ)
Nesogaea (nē-so-jē'-ạ)
Nesomys (nēs'-ô-mis)
Nesophontes (nē-sô-fon'-tēz)
Nesotragus (nē-sot'-rȧ-gus)
Nettion (net'-i-on)
Neuroctena (nûr-ok'-ten-ạ)
neuroglia (nûr-og-lī'-ạ, nûr-ō-glē'-ạ)
neuron (nū'-ron, nū'-rōn)
Neuroptera (nū-rop'-têr-ạ)
Neurotrichus (nū-rot'-rik'-us)
Neviusia* (nev-i-ū'-shi-ạ)
Neyraudia (nā-rô'-di-ạ)
Nezara (nez'-a-rạ)
Nicandra* (nik-an'-drạ)
Nicolletia* (nik-o-le'-ti-ạ)
Nicrophorus (nik-rof'-ôr-us)
nidus (nī'-dus) a nest.
Nierembergia* (nēr-em-bêr'-gi-ạ)
Nigella* (nij-el'-ạ)
niger (nij'-er) black, dark, dusky.
nigrescens (nig-res'-senz) becoming black.
nigricans (nig'-ri-kanz) blackish.
nigritellus (nig-ri-tel'-us) dark, nearly black.
nigritus (nig-rī'-tus) black.
niloticus (nī-lō'-ti-kus) of the River Nile.
nimbosus (nimb-ōs'-us) cloudy, full of rain.
Nirmus (nir'-mus)
Nisaëtus (nis-ā-ē'-tus)
Nisonniades (nis-on-ī'-a-dēz)

nitens (nit'-enz) shining; also, pressing against or upon.

Nitidulidae (nit-i-dū'-li-dē)

nitidus (nit'-i-dus) shining, bright, handsome, rich.

Nitrophila* (nī-trof'-il-ạ)

nivalis (niv-ā'-lis) snowy, belonging to snow.

niveus (niv'-e-us) of or from snow, snowy.

nobilis (nō'-bi-lis) well known, celebrated, noble.

noctiflorus (nok-ti-flō'-rus) flowering at night.

Noctilio (nok-til'-i-ō)

Noctiluca (nok-ti-lū'-kạ)

noctivagans (nok-ti'-va-ganz) night-wandering.

noctivagant (nok-tiv'-ag-ant)

Noctuidae (nok-tū'-i-dē)

nocturnal (nok-tûr'-nal)

nodiflorus (nō-di-flō'-rus) flowering at a node.

Nodosaurus (nō-dô-sô'-rus)

nodose (nōd'-ōs, nō-dōs')

nodosus (nō-dō'-sus) full of knots.

Nolina* (nō-lī'-nạ, nō'-lin-ạ)

Nomada (nom'-a-dạ)

nomenclature (nō-men-klā'-tûr, nō-men'-klà-tûr)

Nomonyx (nō'-mon-iks)

Nonea (non'-ê-ạ)

Nopalea* (nō-pal'-ê-ạ, nō-pā-lē'-ạ)

Nopalxochia* (nō-pal-ksō'-ki-ạ)

nosogenic (nos-ô-jen'-ik)

Nostoc (nos'-tok)

notaeum (nō-tē'-um) pertaining to the back.

Notelaea (not-e-lē'-ạ)

Notemigonus (nō-te-mig-ō'-nus)

Nothofagus (noth-of-āg'-us)

Notholaena* (noth-ol-ē′-nạ)
Notholcus* (noth-ol′-kus)
Nothosaurus (noth-ộ-sộ′-rus)
Nothrotherium (noth-rộ-thē′-ri-um)
Notiosorex (nō-shi-ộ-sō′-reks, nō-ti-ộ-sō′-reks)
Notiothaumidae (nō-shi-ộ-thộ′-mi-dē, nō-ti-ộ-thộ′-mi-dē)
Notodontidae (nō-tộ-don′-ti-dē)
Notogaea (nō-tộ-jē′-ạ)
Notommatidae (nō-tom-at′-i-dē)
Notonectidae (nō-tộ-nek′-ti-dē)
Notophthalmus (nō-top-thal′-mus)
Notoxus (nō-toks′-us)
Notropis (nō′-trộ-pis)
Notungulata (nō-tung-ụ-lā′-tạ)
novenarius (nov-en-ār′-i-us) consisting of or pertaining to the number nine.
nubeculatus (nū-bē-kul-ā′-tus) cloudy, with dark spots.
nubigenus (nū-bi′-jen-us) creating clouds.
nubilus (nū′-bil-us) cloudy, dark, gloomy.
nucellus (nū-sel′-us)
nucha (nū′-kạ)
nuchal (nū′-kal)
Nucifraga (nū-sif′-rạ-gạ)
nucleolar (nū-klē′-ộ-lêr)
nucleolus (nū-klē′-ộl-us)
Nucula (nū′-kụ-lạ)
nudiflorus (nū-di-flō′-rus) with hairless (naked) flowers.
nulliplex (nul′-i-pleks)
Numenius (nū-mēn′-i-us)

Nucifraga, generic name of Clark's Nutcracker <L. *nux*, genit. *nucis*, a nut <*frangere*, to break. Pronounced: nū-sif'-ra-gạ, not nū-si-fra'-gạ.

Nummulites (num-ū-līt'-ēz)
Nuphar* (nū'-fâr)
nuptialis (nup-ti-ā'-lis)
nutans (nū'-tanz) nodding.
Nuttalia* (nut-al'-i-ạ)
Nyctaginia* (nik-tā-jin'-i-ạ)
Nyctale (nik'-ta-lē)
Nyctanassa (nik-tan-as'-ạ)
nyctanthous (nik-tan'-thus)

Nummulites <L. *nummus*, a coin+-*lites* <Gr. *lithos*, a stone. Pronounced: num-ū-lī'-tēz, not nū'-mū-lītz.

Nyctea (nik'-tḕ-ạ)
Nyctereutes (nik-tḕ-rū'-tēz)
Nycteribia (nik-têr-ib'-i-ạ)
Nycteris (nik'-têr-is)
Nycticebus (nik-ti-sē'-bus)
Nycticeius (nik-ti-sē'-i-us)
nyctitropism (nik-tit'-rop-izm), nik-ti-trō'-pizm)
Nyctobates (nik-tob'-at-ēz)
Nyctocalos* (nik-tok'-al-os)
Nymphaea* (nim-fē'-ạ)
nymphaeoides (nim-fē-o-ī'-dēz) like the water-lily.
Nymphalidae (nim-fal'-i-de)
Nyroca (nir-ō'-kạ)
Nysius (nis'-i-us)
Nyssa* (nis'-ạ)

O

Obeliscaria* (ob-el-is-kā'-ri-ạ)
obeliscus (ob-el-is'-kus) an obelisk.
obese (ō-bēs')
obesity (ȯ-bēs'-i-ti, ȯ-bes'-i-ti)
obesus (ȯ-bēs'-us) fat, fattened.
oblique (ob-lēk', ob-līk')
oblongifolius (ob-lon-ji-fol'-i-us, ob-long-ji-fō'-li-
    us), oblong leaf, long leaf.
oblongus (ob-long'-gus) oblong, rather long.
Obolaria* (ob-ȯ-lā'-ri-ạ)
occidentalis (ok-si-den-tā'-lis)
Oceanodroma (ō-shē-an-od'-ro-mạ)
ocellated (os-e-lāt'-ed)
ocellus (ō-sel'-us)
Ochna* (ok'-nạ)

Ochotona <the Tartar name for the pika or little chief-hare, a mammal of
rocky areas of high mountains. Pronounced: ok-ȯ-tō′-nạ.

Ochotona (ok-ȯ-tō′-nạ)
ochraceum (ōk-rā′-se-um) reddish yellow.
Ochranthe* (ō-kran′-thē)
ochroleucus (ō-krȯ-lū′-kus) pale yellow ochre.
Ochroma (ō-krō′-mạ, ok-rō′-mạ)
ochropus (ō-krȯ′-pus) yellow + foot.
Ochrosia* (ō-krō′-si-ạ)
Ochthrodromus (ok-throd′-ro-mus)
Ocimum* (ō′-si-mum, os′-i-mum)
ocrea (ō′-kre-ạ) a legging.
Octadesmia* (ok-tad-es′-mi-ạ)
octomeral (ok-tom′-e-ral)
octopus (ok′-tȯ-pus, pl. ok′-tȯ-pī, also ok-tȯ′-po-
    dēz)
Octopus (ok-tō′-pus)
oculeus (ok-ul′-e-us) full of eyes.
Oculussolis* (ok-ul-us-sō′-lis)
Ocyphaps (ō′-si-faps)
Ocyptera (os-ip′-têr-ạ)
Ocyrhoë (ȯ-sir′-ȯ-ē)

Octopus <L. *octŏpus* <Gr. *oktōpous*, eight-footed. Pronounced: ok-tō'-pus.
The common name "octopus" is accented on the first syllable: ok'-tŏ-pus.

odaks (ō'-daks)
Odinia (ō-din'-i-ạ)
Odobenus (ō-dŏ-bē'-nus)
Odocoileus (od-ŏ-koy'-le-us)
Odonata (ōd-ŏ-nā'-tạ)
Odontarrhena* (od-on-târ'-ren-ạ)
Odontoceridae (od-on-tŏ-ser'-i-dē)
Odontophyes (od-on-tŏ-fī'-ēz)
Odontostomum (ŏ-don-tos'-tŏ-mum)
Odontosyllis (od-on-to-sil'-is)
odoratus (od-ō-rā'-tus) smelling, odorous.
Oecobius (ē-kob'-i-us)
Oedemeridae (ē-dĕ-mer'-i-dē)
Oedicnemus (ē-dik-nē'-mus)
Oedogonium (ēd-ê-gō'-ni-um)
Oenanthe (ē-nan'-thē)
oenocyte (ē'-nŏ-sīt)
Oenothera* (ē-nŏ-thē'-rạ)
Oestrelata (ēs-trel'-ả-tạ)
Oestridae (ēs'-tri-dē)
oestrus (ē'-strus)

officinalis (of-i-si-nā'-lis) of practical use to man, of the apothocary's shop.

Ogcocephalus (og-kŏ-sef'-ȧl-us)

Oidemia (oy-dē'-mi-ạ)

oike (oyk'-ē)

okape (ŏ-kä'-pē)

Okapi (ŏ-kä'-pi)

Okapia (ŏ-kä'-pi-ạ)

Olax* (ol'-aks)

Olea* (ō'-lê-ạ)

Oleaceae (ō-lê-ā'-sê-ē)

Oleacinidae (ō-lê-ȧ-sin'-id-ē)

Oleandra (ō-le-an'-drạ)

Olearia* (ol-e-ā'-ri-ạ)

olecranon (ŏ-le'-krȧ-non)

oleic (ŏ-lē'-ik, ō'-lê-ik)

Oleineae (ō-lê-in'-ê-ē)

Olene (ō-lē'-nē)

olens (ol'-enz) odorous, sweet smelling.

Olenus (ō'-lên-us)

oleraceus (ol-er-ā'-se-us) resembling herbs, vegetable.

Olethreutes (ō-lê-thrū'-tēz)

Olethreutidae (ō-lê-thrū'-ti-dē)

Olfersia (ol-fêr'-si-ạ)

olidus (ol'-i-dus) odorous, of evil smell.

Oligantha* (ol-ig-an'-thạ)

Oligocene (ol'-i-gŏ-sēn)

Oligochaeta (ol-ig-ŏ-kē'-tạ)

Oligomeris* (ol-ig-ŏm'-er-is)

Oligoneuriellidae (ol-ig-ŏ-nūr-i-el'-i-dē)

Oligosma* (ol-ig-oz'-mạ)

Oligotermidae (ol-ig-ŏ-têrm′-i-dē)
olor (ol′-ôr) an odor.
Olusatrum* (ol-us-ā′-trum)
Olyra* (ol-ī′-rạ)
Omalanthus* (om-al-anth′-us)
Omaloptera (om-al-op′-têr-ạ)
Omanus (ŏ-mā′-nus)
ombrophobous (om-brof′-ŏ-bus)
Ommastrephes (om-as′-tre-fēz)
Omosita (ŏm-os-īt′-ạ)
Omphalodes* (om-fal-ŏ′-dēz)
Omus (ŏ′-mus)
onager (on′-ạ-jêr)
Onagra* (ŏ-nā′-grạ)
onca (on′-kạ)
Onchidoris (ong-kid′-ŏ-ris)
Oncidium* (on-sid′-i-um)
Oncifelis (on-sif′-el-is)
Oncocyclus (ong-kos-ī′-klus)
Oncomelania (ong-kŏ-mel-an′-i-ạ)
Oncometopia (ong-kŏ-met-ōp′-i-ạ)
Oncosperma (ong-kos-pêr′-mạ)
oncospheres (ong′-kos-fērz)
Oncotylus (ong-kot′-i-lus)
Ondatra (on-dat′-rạ)
Oniscus (ŏ-nis′-kus)
Onobrychis* (on-ŏb-rī′-kis, on-ob′-rik-is)
Onoclea (on-ok′-le-ạ)
Ononis* (on-ō′-nis)
Onopordon* (on-op-ôr′-don)
Onoseris* (on-os′-er-is)
Onosmodium* (on-os-mō′-di-um)

Onthophagus (on-thof′-ag-us)
Onychium* (on-ik′-i-um)
Onychogalea (on-ik-ŏg-āl′-e-ạ)
Onychomys (on-ik′-ŏ-mis)

MVD—59

Onychomys <Gr. *onyx*, a nail or claw+*mys*, mouse. Generic name of the grasshopper mice. Accent falls on the antepenult. Pronounced: on-ik′ŏ-mis, not on-i-kŏ′-miz as we sometimes hear.

oöcyst (ō′-ŏ-sist)
ooecium (ō-ē′-shi-um, ō-ē′-si-um)
oölogy (ŏ-o′-lŏ-ji)
oötheca (ō-ŏth-ē′-kạ)
operarius (op-er-ā′-ri-us) a workman.
opercular (ŏ-pêr′-kŭ-lâr)
Ophelus (of′-el-us)
Opheodrys (of-ĕ-ōd′-ris)
Ophibolus (of-ib′-ŏ-lus)
Ophidia (of-id′-i-ạ)
Ophiglossum* (of-i-ŏg-los′-um, of-i-ŏ-glō′-sum)
Ophiobolus* (of-i-ob′-ŏ-lus)
Ophiophagus (of-i-ōf′-à-gus)
Ophioplocus (of-i-ŏp-lō′-kus)
Ophiopogon (of-i-ŏ-pō′-gon)
Ophioxylon (of-i-ox-īl′-on)

Ophisaurus (of-i-sô'-rus)
Ophrys* (of'-ris)
Opiliones (op-il-i-ō'-nēz)
Opisthobranchia (op-is-thō-brang'-ki-ạ)
Opisthocomus (op-is-thok'-ō-mus)
opisthotic (op-is-tho'-tik)
Oplismenus* (op-lis'-men-us)
Opomiza (op-ō-mī'-zạ)
Opopanax* (op-op'-an-aks, ô-pop'-à-naks)
Oporanthus* (op-ôr-an'-thus)
Oporornis (op-ôr-ôr'-nis)
Opostega (op-os'-te-gạ)
Opsebius (op-sē'-bi-us)
opthalmic (op-thal'-mik)
Opuntia (ō-pun'-shi-ạ, ō-pun'-ti-ạ, op-un'-ti-ạ)
orarius (ō-rā'-ri-us) of or belonging to the coast.
Orasema (ôr-as-ēm'-ạ)
orbicularis (ôr-bik-u-lā'-ris) circular, in the shape
    of an orb.
Orca (ôr'-kạ)
Orchestes* (ôrk-es'-tēz)
Orchis* (ôr'-kis)
Orcinus (ôr-sī'-nus)
Ordovician (ôr-dō-vish'-i-an)
ordure (ôr'-dûr)
Oreamnos (ô-rē-am'-nos)
Orelia* (ôr-el'-i-ạ)
Oreocharis* (ôr-e-ok'-âr-is)
Oreodaphne* (ôr-e-od'-af-nē)
Oreohelix (ôr-e-o'-hel-iks)
Oreoscoptes (ôr-e-ô-skōp'-tēz, ôr-e-ô-skop'-tēz)
Oreotragus (ô-re-ot'-rạ-gus)

Oreta (ôr-ēt′-ạ)
orientalis (ô-ri-en-tā′-lis) belonging to oriens, the
   East.
Origanum* (ôr-ī′-gan-um, ō-rig′-a-num)
originalis (ō-ri-ji-nā′-lis) primitive, original.
oriundus (ôr-i-un′-dus) descended, sprung from.
orius (ôr′-i-us) mountain-dwelling, mountain.
Ormenis (ôr′-men-is)
Ormyrus (ôr-mī′-rus)
ornatulus (ôr-nā′-tu-lus) fine, smart.
Orneodes (ôr-ne-ōd′-ēz)
Ornithogalum* (ôr-ni-thog′-al-um)
Ornitholestes (ôr-nith-ō-les′-tēz)
ornithology (ôr-ni-thol′-ōj-i)
Ornithopus* (ôr-nith′-op-us, ôr-nī′-thop-us)
Orobanche* (ôr-ob-ang′-kē)
Orobella* (ôr-ob-el′-ạ)
Orobus* (ôr′-ob-us)
Orohippus (ôr-ŏ-hip′-us)
orolestes (ôr-ŏ-lēs′-tēz) a mountain-robber.
Orontium* (ôr-on′-shi-um, ŏ-ron′-ti-um)
Oroxylum* (ôr-oks′-il-um)
Ortalis (ôr′-tạ-lis)
Orthezia (ôrth-ēz′-i-ạ)
Orthocarpus* (ôr-thŏ-kâr′-pus)
Orthocladius (ôr-thŏ-klad′-i-us)
Orthogeomys (ôr-thŏ-jē′-ŏ-mis)
Orthonyx (ôr′-thŏ-niks)
Orthoptera (ôr-thop′-têr-ạ)
Orthotomus (ôr-thot′-ŏ-mus)
ortus (ôr′-tus) sprung from, descended.
Ortygometra (ôr-ti-gŏ-mē′-trạ)

Ortygospiza (ôr-ti-gŏ-spī'-zạ)
Orussidae (ŏ-rus'-i-dē)
Orycteropus (ôr-ik-ter'-ŏ-pus)
Oryctes (ôr-ik'-tēz)
Oryctolagus (ôr-ik-tol'-a-gus)
Oryssus (ŏ-ri'-sus)
Oryx (ō'-riks, ôr'-iks)
Oryza (ŏ'-rī'-zạ)
Oryzomys (ôr-i'-zŏ-mis, ôr-ī'-zŏ-mis)
Oryzopsis* (ôr-i-zop'-sis, ôr-ī-zop'-sis)
Oscinis (os'-i-nis)
osmeterium (os-me-tē'-ri-um)
Osmorrhiza* (os-mô-rī'-zạ)
osmosis (os-mō'-sis, oz'-mō-sis)
osmotic (os-mot'-ik)
Osmunda* (os-mun'-dạ)
Osmylidae (os-mi'-li-dē)
osphradium (os-frā'-di-um)
Osphranter (os-fran'-têr)
osphretic (os-frēt'-ik)
osphresis (os-frē'-sis)
osprey (os'-prā, os'-pri)
Osteolaemus (os-te-ŏ-lē'-mus)
Osteospermun* (os-te-os-pêr'-mum)
Ostinops (os'-ti-nops)
Ostomatidae (os-tŏ-mat'-i-dē)
Ostracoda (os-trȧ-kō'-dạ, os-trak'-ŏ-dạ)
Ostracoderm (os'-trȧ-kŏ-dêrm, os-trak'-ŏ-dêrm)
Ostrea (os'-trē-ạ)
ostreatus (os-tre-ā'-tus) rough, scabby.
Ostruthium* (os-trū'-thi-um)
Ostriya* (os'-tri-ạ)

Osyris* (os′-ir-is)
Otaria (ȯ-tā′-ri-ạ)
Othnius (oth′-ni-us, oth-nī′-us)
Othonna* (ō-thon′-ạ)
Otides (ō′-ti-dēz)
otidium (ȯ-tid′-i-um)
Otis (ō′-tis)
Otocorys (ȯ-tok′-ȯ-ris)

Otocorys <Gr. *ous* (ōt), ear+*korys*, helmet. Also spelled Otocoris. The genus
includes the horned larks. Pronounced: ȯ-tok′-ȯ-ris, not ōt-ō-kȯr′-is.

Otocyon (ȯ-tos′-i-on)
Ototylomys (ōt-ȯ-tī′-lȯ-mis)
Otus (ō′-tus)
ovatus (ō-vā′-tus) egg-shaped; also, having egg-
    shaped spots.
ovinus (ov-ī′-nus) belonging to sheep.
oviparous (ō-vi′-pa-rus)
Ovis (ō′-vis)
Oviscapte (ō-vis-kapt′-ē)
ovule (ō′-vūl)
Oxalis* (ok′-sa-lis)
Oxybaphus* (oks-ib′-ạ-fus)

Oxalis. New Latin. <Gr. *oxys*, acid.
Pronounced: oks'-al-is, not oks-
al'-is.

Oxybelis (oks-ib'-el-is)
Oxycoccus* (oks-i-kok'-us)
Oxydendrum* (oks-id-en'-drum)
Oxyechus (oks-i-ē'-kus)
Oxyopes (oks-i-ō'-pēz)
oxyphilous (oks-if'-i-lus)
Oxypoda (oks-ip'-ŏ-dạ)
Oxyptilus (oks-ip'-til-us)
Oxyria* (oks-ir'-i-ạ)
Oxyropus* (oks-ir'-ŏ-pus)
Oxystylis* (oks-i-stī'-lis)
Oxytelus (oks-it'-ê-lus)
Oxytenia* (oks-it-ē'-ni-ạ)
Oxytropis* (oks-it'-rop-is, oks-it'-rŏ-pis)
Ozaena (ô-zēn'-ạ)
Ozothamnus* (oz-oth-am'-nus)

P

Pachidendron* (pak-id-en'-dron)
Pachira* (pak-ī'-rạ)

Pachistima* (pak-is'-ti-mạ)
Pachybrachys (pak-ib'-rak-is)
Pachycereus* (pak-i-sē'-rê-us)
Pachycormis* (pak-i-kôr'-mis)
Pachygrapsus (pak-i-grap'-sus)
Pachylomerides (pak-i-lō-mer'-ī-dēz)
Pachypoda (pak-ip'-ŏ-dạ)
Pachyrhizus* (pak-i-rī'-zus)
Pachysandra* (pak-is-an'-drạ)
Pachystima* (pak-is'-ti-mạ)
Pachystoma* (pak-is'-tom-ạ)
pademelon (pad'-ê-mel-on)
Paederia* (pē-dē'-ri-ạ)
Paederus (pē'-der-us)
paedogenesis (pē-dŏ-jen'-e-sis)
Paeonia* (pê-ō'-ni-ạ)
Pagasa (pa'-ga-sạ)
Pagina* (pā'-jin-ạ)
Pagiopoda (pā-ji-op'-ŏ-dạ)
Pagolla (pag-ol'-ạ)
Pagomys (pag'-ŏ-mis)
Pagophila (pag-of'-i-lạ)
Paguma (pȧ-gū'-mạ)
Pagurus (pȧ-gū'-rus)
Piaropus* (pī-ar'-ŏ-pus)
paisano (pī-sä'-nō)
palaearctic (pā-lê-ârk'-tik)
Palaemon (pȧ-lē'-mon)
Palaeochenoides (pā-lê-ŏ-kēn-o-ī'-dēz)
Palaeoscincus (pā-lê-ŏ-skink'-us)
Palafoxia* (pä-läf-ok'-si-ạ)
Palamedea (pal-ȧ-mē'-dê-ạ)

palea (pā'-lê-ạ)
paleaceus (pal-e-ā'-se-us) like chaff, chaffy.
Paleacrita (pāl-ê-ak'-ri-tạ, pal-ê-ak'-ri-tạ)
paleobotany (pā-lē-ô-bot'-ạ-ni, pal-e-ô-bot'-ạ-ni)
Paleolaria* (pā-le-ol-a'-ri-ạ, pal-e-ol-a'-ri-ạ)
paleolithic (pā-lê-ô-lith'-ik, pal-ê-ô-lith'-ik)
paleophytic (pā-lê-ô-fit'-ik, pal-ê-ô-fit'-ik)
Paleozoic (pā-le-ô-zô'-ik, pal-ē-o-zo'-ik)
palingenesis (pal-in-jen'-e-sis)
Palingeniidae (pal-in-jen-ī'-i-dē)
Palinurus (pal-i-nū'-rus)
Paliurus* (pal-i-ū'-rus)
Pallavicinia* (pal-av-i-si'-ni-ạ)
pallescens (pal-es'-senz) turning pale.
palliatus (pal-i-ā'-tus)
pallidus (pal'-i-dus) pale.
pallium (pal'-i-um)
palmatisect (pal-mat'-i-sekt)
palpebra (pal'-pê-brạ)
palpebral (pal'-pe-bral)
Paltonium* (pal-ton'-i-um)
Paludicolae (pal-ū-dik'-ô-lē)
Paludina (pal-û-dī'-nạ)
paludinal (pal-ū'-di-nal)
paludose (pal'-û-dōs)
paludosus (pal-û-dō'-sus) marshy, boggy.
Palumbina* (pal-um-bī'-nạ)
palus (pā'-lus, pl. pā'-lī)
paluster (pal-us'-ter) swampy, marshy.
palynology (pal-in-ol'-jô-i)
Pamphila (pam'-fi-lạ)
Pamphiliidae (pam-fil-ī'-i-dē)

Panagaeus (pan-a-jē'-us)
pancreas (pan'-krē-as)
Pandaca* (pan'-dak-ą)
Pandanus* (pan'-dan-us, pan-dā'-nus)
Pandion (pan-dī'-on, pan'-di-ōn)

Pandion <Gr. *Pandion* >L.
Pandion, king of Athens, father
of Procne, supposed to have
been changed into a swallow.
Pronounced: pan-dī'-on, not
pan'-di-on.

Pandorea* (pan-dō'-rĕ-ą)
Pandorina (pan-dŏ-rī'-ną)
pangamic (pan-gam'-ik)
Pangaeus (pan-jē'-us)
pangens (pan'-jenz)
paniceus (pā-ni'-se-us) made of bread.
paniculatus (pan-i-kŭ-lā'-tus) having pannicles or
   tufts of flowers.
Panicum* (pan'-i-kum)
pannosus (pan-ōs'-us) full of rags.
Panorpidae (pan-ôr'-pi-dē)
panthalassic (pan-thal-as'-ik)
Pantoclis (pan'-tok-lis)
Panulirus (pan-ūl'-ir-us)
Papaver* (pă-pā'-vêr, pap-ā'-vêr)
Papaya* (pa-pī'-ą)
Paphia (pā'-fi-ą)

Paphiopedilum* (pā-fi-ȯ-ped'-i-lum)
Papirius (pap-īr'-i-us)
papillary (pap'-i-lā'-ri, pȧ-pil'-ȧ-ri)
Papio (pā'-pi-ō)

Papio <Fr. *papion*, the baboon. Pronounced: pa'-pi-o.

Pappogeomys (pap-ȯ-jē'-o-mis)
Pappophorum* (pap-of'-ôr-um)
pappus (pap'-us)
papyraceus (pap-ī-rā'-se-us) papery.
parabiosis (par-ȧ-bī-ōs'-is)
Paracaryum* (par-ak-ar'-i-um)
Paracrangon (par-ȧ-kran'-gon)
Paracyamus (par-ȧ-sī'-a-mus)
Paradisia* (par-ad-ī'-si-ạ)
paradisiaca (par-ad-ī-si'-ak-ạ)
Paradoxurus (par-ȧ-dok-sū'-rus)
paradoxus (par-ȧ-doks'-us) strange, contrary to
   expectation.
Paragalia (par-ag-ā'-li-ạ)
Paragramma* (par-ag-ram'-ạ)
Paragus (par'-ȧ-gus)

Parahippus (par-à-hip′-us)
Paralariscus (par-al-âr-isk′-us)
Paralichthys (par-à-lik′-this)
paralius (par-al′-i-us) that grows by the seaside.
Parameles (par-am′-ē-lēz)
Paramesius (par-am-ē′-si-us)
Parandra (par-an′-drą)
Parapholas (par-af′-ŏ-las)
paraphysis (par-af′-i-sis)
parapodium (par-à-pō′-di-um)
parapsidal (par-ap′-si-dal)
parapsis (par-ap′-sis)
Parascalops (par-as′-kal-ops)
parasitism (par′-ą-sīt-izm)
Parastacus (par-as′-tå-kus)
Pardalianches* (pâr-dal-i-ang′-kēz)
paradalis (pâr′-da-lis) a female panther, also, a tiger.
Pardanthus* (pâr-dan′-thus)
Pardalotus (pâr-dą-lōt′-us)
Pareiasauria (par-ē-à-sŏ′-ri-ą)
Paridra (par′-i-dra, par-ī′-drą)
paries (pa-rī′-ēz, pl. pa-rī′-et-ēz)
Parietaria* (par-i-et-ā′-ri-ą, pà-rī-e-tā′-ri-ą)
parietes (pa-rī′-et-ēz)
paris (par′-is) equal.
Parnassia* (pâr-nas′-i-ą)
Parnassiidae (pâr-nas-ī′-i-dē)
Parnassius (pâr-nas′-i-us)
Parnopes (pâr-nō′-pēz)
Paronychia* (par-ŏ-nik′-i-ą)
Parosela* (par-ŏ-sē′-lą)

parotid (på-rō'-tid, par-ot'-id)
Parthenium* (pâr-the'-ni-um)
parthenogenesis (pâr-then-ō-jen'-e-sis)
Parula (pâr'-ủ-lạ)
Parus (pā'-rus)
parvifolus (pâr-vi-fol'-i-us, pâr-vi-fō'-li-us) with
    small leaves.
parvulus (pâr'-vu-lus) very small, slight.
Pasimachus (på-sim'-ạ-kus)
Passerculus (pas-êr'-kủ-lus)
Passeres (pas'-êr-ēz)
Passerherbulus (pas-êr-erb'-ủ-lus)
Passerina (pas-êr-ī'-nạ)
passerinus (pas-er-ī'-nus) like a sparrow.
Passiflora* (pas-i-flō'-rạ)
Pastinaca* (pas-tin-ā'-kạ)
patagium (pat-a'-ji-um, pat-ā'-ji-um)
Patamon (pat'-ạ-mon)
patens (pat'-enz) open, accessible.
patent (pā-tent, pat'-ent)
patina (pat'-in-ạ)
Patriofelis (pā-tri-ȯ-fēl'-is)
patruelis (pat-ru-el'-is) a cousin.
patulus (pat'-u-lus) open, spread out, broad; also,
    common.
pauciflorus (pȯ-si-flō'-rus) with few flowers.
paulus (pȯ'-lus) small.
paunch (pänch, pȯnch)
Paurotes (pȯr-ō'-tēz)
Paurotis (pȯr-ō'-tis)
Pauxi (pȯk'-si)
Pavo (pā'-vō)

Pavo <L. *pāvo*, the peacock. Pronounced: pā'-vō, not pa'-vō.

Pavonaria (pā-vŏ-nā'-ri-ạ)
Pavonia* (pā-vō'-ni-ạ)
paxilla (pak-sil'-ạ)
pebrine (pe-brēn', pe'-brin)
pecan (pē-kän', pĕ-kan')
pectineal (pek-tin'-e-al)
pectoralis (pek-tŏ-rā'-lis)
pedalis (ped-ā'-lis) of or belonging to the foot, a
　　foot in length; also, a slipper.
Pedetes (pē-dē'-tēz)
Pedetidae (pē-det'-i-dē)
pedicellaria (ped-i-sel-ā'-ri-ạ)
Pedicularis* (ped-ik-ū-lā'-ris)
Pediculidae (ped-i-kū'-li-dē)
Pedilanthus* (ped-i-lan'-thus)
Pedilonum* (ped-i-lō'-num)
Pedilus (ped'-il-us)
Pedioecetes (ped-i-ŏ-sē'-tēz)
Pedionomus (ped-i-on'-ŏ-mus)
Pedipes (ped'-i-pēz)
pedonic (ped-on'-ik)
Peganum (pē'-gan-um, peg'-an-um)

pelage (pel'-aj)
pelagicus (pel-a'-ji-kus) relating to the sea.
Pelargonium* (pel-âr-gō'-ni-um)
Pelecanus (pel-e-kā'-nus)
Pelecinus (pel-es-īn'-us)
Pelecypoda (pel-e-si'-pŏ-dạ)
pelegrina (pel-e-grī'-nạ)
Pelidna (pel-id'-nạ)
Pelidnota (pel-id-nō'-tạ)
pelius (pel'-i-us) black, livid.
Pellaea* (pel-ē'-ạ)
pellions (pel'-i-onz)
pellucidus (pel-ū'-si-dus) transparent.
Pelobates (pē-lob'-à-tēz)
Pelocoris (pel-ok'-ôr-is)
Pelopaeus (pel-ŏ-pē'-us)
peloria (pel-ō'-ri-ạ)
pelta (pel'-tạ) a half-moon shaped shield.
peltatus (pel-tā'-tus) having shields.
pelvis (pel'-vis, pl. pel'-vēz)
Pempheris (pem-fē'-ris)
Pemphredonidae (pem-frê-don'-i-dē)
pendulus (pen'-du-lus) hanging, pendent; also, doubtful.
Peneides (pen-ē-ī'-dēz)
penelope (pē-ne-lo'-pē)
penicillatus (pē-nis-il-ā'-tus)
penis (pē'-nis, pl. pē'-nēz)
Pennisetum* (pen-is-ē'-tum)
pennus (pen'-us) pointed, sharp; also, a wing.
Pentachaeta* (pent-ak-ē'-tạ)
Pentacrinus (pen-tak'-ri-nus, pen-tak-rī'-nus)

Pentapetes* (pent-ap′-et-ēz)
Pentaptera* (pent-ap′-te-rạ)
Pentarthron (pent-âr′-thron)
Pentatoma (pent-at′-ọ̇-mạ)
Pentatomidae (pent-a-tom′-i-dē)
Penthestes (pen-thes′-tēz)
Penthina (pen-thī′-nạ)
Penthorum* (pen′-thọ̇-rum)
Pentstemon* (pent-stē′-mon)
Peperomia* (pep-êr-ō′-mi-ạ)
peplis (pep′-lis) the name of some plant.
peploides (pep-lo-i′-dēz) like *Peplis*.
pepo (pē′-pō, pep′ō)
Peraclius (per-ak-lī′-us)
Perdicium* (pêr-dī′-si-um, pêr-dī′-shi-um)
perditus (pêr′-di-tus) ruined, made away with.
Perdix (pêr′-diks)
peregrine (per′-ê̇-grin)
peregrinus (per-e-grīn′-us) strange, foreign.
perennis (per-en′-is) continuing through the year,
    unfailing.
Pereskia* (per-esk′-i-ạ)
Perezia* (pē-rē′-zi-ạ)
perfoliatus (per-fol-i-āt′-us) having the stems
    appearing to pass through a leaf.
perforatus (per-for-ā′-tus) piercing through.
Pericallis* (per-ik-al′-is)
Perichaena* (per-i-kē′-nạ)
periclinal (per-i-klī′-nal)
Pericome* (per-ik′-ọ̇-mē)
pericranial (per-i-krā′-ni-al)
Peridinium (per-i-din′-i-um)

Peridroma (per-id'-rom-ą)
Perigonimus (per-i-gōn'-i-mus)
perigonium (per-i-gōn'-i-um)
perigynous (per-ij'-in-us)
Perilampidae (per-i-lamp'-i-dē)
Perilla* (pĕ-ril'-ą)
Perillus (pĕ-ril'-us)
Periophthalmus (per-i-of-thal'-mus)

Peripatus. Pronounced: per-ip'-a-tus, not per-i-pā'-tus.

Peripatus (per-ip'-ă-tus)
periphery (per-if'-êr-i)
periphloic (per-i-flō'-ik)
periphysis (per-if'-is-is)
Periplaneta (per-i-plăn-ē'-tą)
Periploca* (per-ip'-lŏk-ą)
Perisoreus (per-i-sō'-re-us)
peristalsis (per-i-stal'-sis)
Peristeria* (per-is-tē'-ri-ą)
Peritoma (per-it'-ŏ-mą)
peritoneum (per-i-tŏ-nē'-um)
peritrichous (per-it'-ri-kus)
Perityle* (per-i'-ti-lē)

Perlidae (pêr'-li-dē)
Pernettya* (pêr-ne'-ti-ạ)
Perognathus (pē-rog'-na-thus)
Permian (pêr-mi'-an)
Peromya (pēr-ȯ-mī'-ạ)
Peromyscus (pēr-ȯ-mis'-kus)

Perognathus. The Spiny Pocket Mouse <Gr. *pĕra*, pouch+*gnathos*, jaw
Pronounced: pĕ-rog'-nả-thus, not per-ȯ-gnả-thus.

Peronospora* (per-ȯ-nos'-pȯ-rạ)
peropodous (pê-rop'-ȯ-dus)
perpinguis (per-pin'-gu-is) very rich.
Perrisonetta (per-is-ȯ-net'-ạ)
Persea* (pêr-sē'-ạ)
persicifolius (pêr-si-ki-fol'-i-us, pêr-si-ki-fō'-li-us)
   with leaves like the peach.
personus (pêr'-son-us) ringing, resounding.
pertinax (pêr'-ti-naks) tenacious, obstinate, persistent.
pertusus (pêr-tū'-sus) perforated.
perulate (per'-u-lāt)
pes (pēz, pl. pē'-dēz)
pessulus (pes'-ů-lus)
Petalostemon* (pet-al-os-tē'-mon)

Petasites* (pet-à-sī′-tēz)
Petaurista (pet-ô-ris′-tą)
petax (pet′-aks) greedy.
petilus (pet-ī′-lus) thin, slender.
petiole (pet′-i-ōl)
petraeus (pet-rē′-us) growing among rocks.
Petrea* (pet-rē′-ą)
petrel (pet′-rel) little Peter.
Petricola (pē-trik′-ô-lą)
Petrochelidon (pet-rô-kel′-i-don)
Petrogale (pet-rog′-à-lē)
Petrophila* (pet-rof′-il-ą)
Petroselinum* (pet-ros-el-ī′-num)
petrous (pet′-rus)
petunia (pet-ū′-ni-ą)
Peucaea (pū-sē′-ą)
Peucedanum* (pū-sed′-à-num)
Peucedramus (pū-sed′-ram-us)
Peucephyllum* (pū-se-fil′-um)
peyote (på-yō′-tê, på-yō′-tå)

Peucephyllum <Gr. *peukē*, the pine or fir
+*phyllon*, leaf. Pronounced: pū-sē-fil′-um.

Peziza* (pė-zī'-zą, pez-iz'-ą)
Pezophaps (pez'-ŏ-faps)
Phacelia* (fȧ-sē'-li-ą)
Phacochoerus (fak-ŏ-kē'-rus)
phacoid (fāk'-oid, fak'-oid)
Phaedon* (fē'-don)
Phaedranassa* (fēd-ran-as'-ą)
Phaedranthus* (fēd-ran'-thus)
phaeism (fē'-izm)
phaeocryptus (fē-ŏ-krip'-tus) dusky + hidden.
Phaeophycophyta (fē-ŏ-fī-kof'-it-ą)
Phaeopus (fē'-ŏ-pus)
Phaëthon (fā'-e-thon)
phage (fāj)
phagocyte (fag'-ŏ-sīt)
phagolysis (fag-ol'-is-is)
Phainopepla (fā-i-nŏ-pep'-lą)
Phajus* (fā'-jus)
Phalacrocorax (fal-a-krō'-kŏ-raks)
Phalaenopsis* (fal-ė-nop'-sis)
Phalaenoptilus (fal-ė-nop'-til-us)

Phainopepla   <Gr.
phaeinos shining+*pep-
los*, a robe. Pronounced:
fā-i-nŏ-pep'-lą, fȧ-i-nŏ-
pēp'-lą.

phalanger (fă-lan'-jêr)
phalanx (fal'-angks, pl. fal-an'-jēz)
Phalaris* (fal-âr'-is)
phalarope (fal'-à-rōp)
Phalaropus (fal-âr'-ŏ-pus)
phallus (fal'-us)
Phalonia (fal-ōn'-i-ạ)
Phanaeus (fan-ē'-us)
Phaneroglossus (fan-er-ŏ-glos'-us)
phaosome (fā'-ŏ-sōm)
Pharbitis* (fâr-bī'-tis)
Pharomacrus (fâr-ŏ-mak'-rus)
Pharus (fā'-rus)
Phascogale (fas-kog'-al-ē)
Phascolarctos (fas-kŏ-lârk'-tos)
Phascolomus (fas-kol'-ŏ-mus)
Phascolomys (fas-kol'-ŏ-mis)
Phaseolus* (fā-sē'-ŏ-lus, fas-ē'-ŏ-lus)
Phasianus (fā-si-ā'-nus)
Phasmida (faz'-mi-dạ)
Phasmidae (faz'-mi-dē)
Phataginus (fat-a'-ji-nus)
Phebalium* (feb-al'-i-um)
Phegopteris* (fĕ-gop'-ter-is)
Pheidole (fī-dō'-lē)
phellema (fel-ē'-mạ)
Phenacodus (fen-ak'-o-dus)
Phenacomys (fen-ak'-ŏ-mis)
phengophobe (feng'-ŏ-fōb)
phenogamous (fen-og'-am-us)
phenol (fē'-nōl, fen'-ol)
phenotype (fēn'-ŏ-tīp, fen'-ŏ-tīp)

phialide (fī'-al-īd)
Phidippus (fī-di'-pus)
Philacte (fil-ak'-tē)
Philander (fil-an'-dêr)
Philetaerus (fil-ê-tē'-rus)
Philohela (fil-o'-he-lạ)
Philomachus (fil-om'-à-kus)
Philonthus (fil-on'-thus)
Philopteridae (fil-op-ter'-i-dē)
Phlebodium* (fleb-ō'-di-um)
Phlebotomus (fleb-ot'-ô-mus)
Phlegethontius (fleg-eth-on'-ti-us)
Phleum* (flē'-um)
phloem (flō'-em)
phloeoterma (flē-ot-êr'-mạ)
Phloeophora (flé-of'-ôr-ạ)
Phloeothripidae (flē-ô-thrip'-i-dē)
Phloeotomus (flē-ot'-ô-mus)
Phlogacanthus* (flog-ak-an'-thus)
Phlomis* (flō'-mis, flom'-is)
Phoca (fō'-kạ)
Phocaena (fô-sē'-nạ)
Phoenicopterus (fē-ni-kop'-têr-us)
Pholadidea (fō-lad'-i-dē)
Pholidauges (fol-id'-ôj-ēz)
Pholcus (fol'-kus)
Pholidota (fol-i-dō'-tạ)
Pholisma* (fol-iz'-mạ)
Pholistoma* (fol-is-tō'-mạ)
Pholiurus* (fol-i-ūr'-us)
Phora (fō'-rạ)
Phoradendron* (fō-rà-den'-dron)

Phorantha (fō-ran'-thạ)
Phoridae (fōr'-i-dē)
Phormium* (fôr'-mi-um)
Phorodon (fôr-ōd'-on)
Photinus (fō-tīn'-us)
phototropism (fō-tot'-rō-pizm)
phoxocephalus (foks-ō-se'-fal-us) tapering head.
Phragmatobia* (frag-mat-ob'-i-ạ)
Phragmites* (frag-mī'-tēz)
phragmocyttarous (frag-mō-sit'-âr-us)
phratry (frā'-tri)
phreneticus (fren-ē'-ti-kus) mad, delirious.
Phrixocephalus (frik-sō-sef'-al-us)
Phryganeidae (frig-ạ-nē'-i-dē)
Phryma (frī'-mạ)
Phrynichus (frin'-i-kus)
Phrynium* (frī'-ni-um)
Phrynosoma (frī-no-sō'-mạ)
Phthiridae (thir'-i-dē)
Phthirius (thir'-i-us)
Phyciodes (fis-ī'-ō-dēz)
Phycis (fī'-sis)
Phycita (fī'-sit-ạ)
Phycomycetes (fī-ko-mī-se'-tēz)
Phylachora (fī-lak'-ō-rạ)
phylicifolus (fi-li-si-fol'-i-us, fi-li-si-fō'-li-us) with
    leaves like *Phylica*.
Phyllanthus* (fil-an'-thus)
phyllary (fil'-ạ-ri)
Phyllidae (fil'-i-dē)
phylloclade (fil'-ō-klād)
Phyllocnistis (fil-ok-nis'-tis)

Phyllodactylus (fil-lŏ-dak′-ti-lus)
phyllode (fil′-ōd)
Phyllodoce* (fil-od′-ŏ-sē)
Phyllodromiidae (fil-ŏ-drom-ī′-i-dē)
Phyllomedusa (fil-ŏ-mĕ-dū′-sạ)
Phyllopoda (fil-op′-ŏ-dạ)
Phyllorynchus (fil-ŏ-ring′-kus)
Phylloscopus (fil-os′-kŏ-pus)
Phyllostachys* (fil-os′-tā-kis)
Phyllotreta (fil-ŏ-trē′-tạ)
Phylloxera (fil-ok-sē′-rạ)
Phylloxeridae (fil-ok-ser′-i-dē)
phylogeny (fī-loj′-ĕ-ni)
Phymata (fī′-mat-ạ)
Phymatidae (fī-mat′-i-dē)
Phyostegia* (fī-ŏ-stej′-i-ạ)
Physalia (fī-sā′-li-ạ)
Physalis* (fī′-sal-is)
physalus (fī′-sa-lus) the rorqual whale.
Physcia* (fis′-i-ạ)

Physalis. New L. <Gr. *Phȳsa*, a bladder, a bel-
lows. Pronounced: fī′-sal-is, not fis′-al-is.

Physeter (fī-sē'-tēr)
Physianthus* (fī-si-an'-thus)
Physocarpus* (fī-sŏ-kâr'-pus)
Physosiphon* (fī-sos'-if-on)
Physospermum* (fī-sos-pêrm'-um)
Physostegia* (fī-sos-tē-ji'-ạ, fī-sŏ-ste'-ji-ạ)
Phytelephas* (fī-tel'-ê-fas)
Phyteuma* (fit-ū'-mạ)
phytome (fī'-tōm)
Phytophaga (fī-tof'-à-gạ)
Pica (pī'-kạ)
Picea* (pīs'-ê-ạ)
pichiciago (pich-i-si-ä'-gō)
Pici (pī'-sī)
Picicorvus (pis-i-kôr'-vus, pī-si-kôr'-vus)
Picinae (pis-ī'-nē)
Picoides (pik-o-i'-dēz)
Picramnia* (pik-ram'-ni-ạ)
Picris* (pik'-ris)
pictus (pik'-tus) painted, stained.
Picumnus (pi-kum'-nus)
Picus (pī'-kus)
Pieris (pī'-er-is)
Piesma (pī-ēs'-mạ)
pigal (pī'-gal)
pigidium (pī-jid'-i-um)
pika (pī'-kạ)
Pilea* (pī'-le-ạ)
pileate (pī'-lê-at, pil'-ê-ât)
pileatus (pī-le-ā'-tus) capped.
pileolus (pil-ē'-ŏl-us)
pilidium (pī-lid'-i-um)

pilifer (pil'-if-êr)
pilomotor (pil-ộ-mōt'-ôr)
pilose (pil'-ōs)
pilosus (pil-ō'-sus) shaggy, hairy.
Pilularia* (pil-ul-ā'-ri-ạ)
pilulifera (pil-ul-if'-er-ạ) bearing small pill-like
    glands.
Pimelea* (pī-mel'-e-ạ)
Pimephales (pim-ef-ā'-lēz)
Pimpinella* (pim-pin-el'-ạ)
pimpinelloides (pim-pin-el-o-ī'-dēz) pimpernel-like.
Pinanga* (pin-ang'-ạ)
pineal (pī'-ne-al, pin'-e-al)
pinetum (pīn-ē'-tum)
Pinguicula (pin-gwik'-û-lạ)
Pinnipedes (pin-ip'-ê-dēz)
Pinnipedia (pin-i-pē'-di-ạ)
Pinnotheres (pin-ộ-thē'-rēz)
pinocytosis (pin-os-īt-ōs'-is)
Pinus* (pī'-nus)
Piophilidae (pī-ộ-fil'-i-dē)
Pipa (pī'-pạ)
Piper* (pī'-per, pip'-er)
piperatus (pi-per-ā'-tus) with peppery taste.
pipiens (pī'-pi-enz) chirping.
Pipile (pip-ī'-lē)
Pipilo (pip'-i-lō)
Pipistrellus (pip-is-trel'-us)
Pipridae (pip'-ri-dē)
Piptochaetium* (pip-tō-kē'-shi-um, pip-tộ-kē'-ti-
    um)
Pipunculidae (pi-pung-kū'-li-dē)

Piranga (pī-rang′-ą)
pisces (pī′-sēz)
piscine (pis′-īn, pis′-in)
Pisidae (pis′-i-dē)
Pisidium (pis-id′-i-um)
pisiform (pis′-i-fôrm)
pisiformis (pis-i-fôrm′-is) pea-form.
Pisobia (pis-ō′-bi-ą)
Pissodes (pis-ō′-dēz)
Pistacia* (pis-tā′-shi-ą)
pitahaya (pē-tä-hä′-yä)
Pithecanthropus (pith-ê-kan-thrō′-pus)
Pithecolobium* (pith-ê-kŏl-ōb′-bi-um)
Pithium* (pith′-i-um)
Pittosporum* (pit-os′-pô-rum)
Pituophis (pit-û-ō′-fis)
Pitymys (pit′-i-mis)
Pityophis (pit-ī′-ô-fis)
Pityrogramma (pit-ī-rô-gram′-ą)
Pizonyz (pīz′-on-iks)
Placea* (plas′-e-ą)
Plagiobothrys*(plā-ji-ô-bo′-thriz, plaj-i-ô-bo′-thris)
Plagiodon (plā-ji′-ô-don, plaj′-i-ô-don)
plancus (plan′-kus) a kind of eagle.
Plantago* (plan-tā′-gō)
planula (plan′-ū-lą)
Plasmodium (plaz-mō′-di-um)
Platalea (plat-ā′-lê-ą)
Plataleidae (plat-à-lē′-i-dē)
Platanus* (plat′-à-nus)
Platax (plā′-taks)
Platycerus (plat-is′-er-us)

Plantago <L. *plantago*, the plantain. Pronounced: plan-tā'-gō, not plan-tä'-gō, unless you pronounce it according to the Roman method.

Platyclinis* (plat-ik-lī'-nis)
Platycodon* (plat-i-kō'-don)
Platyctenea (plat-i-te'-ne-ạ)
Platydema (plat-id'-em-ạ)
Platygeomys (plat-i-gē'-ŏ-mis)
Platyhelminthes (plat-i-hel-min'-thēz)
Platypeza (plat-ip-ēz'-ạ)
Platypezidae (plat-i-pez'-i-dē)
platyphyllus (plat-i-fil'-us) flat leafed.
Platypsaris (plat-ip'-sȧ-ris)
Platypsyllus (plat-ip-sil'-us)
Platypteris* (plat-ip'-ter-is)
Platyptilia (plat-ip-til'-i-ạ)
Platysomus (plat-i-sō'-mus)
Platyspermum* (plat-i-spêr'-mum)
Plecia (plē'-si-ạ)
Plecoptera (plē-kop'-têr-ạ)
Plecotus (plē̆-kō'-tus)
Plectocomia* (plek-tok-om'-i-ạ)
Plectrophenax (plek-trof'-e-naks)
Plegadis (plē'-ga-dis, pleg'-a-dis)

Pleioblastus* (plī-ŏ-blast'-us)
pleiogonus (plī-og'-on-us) many-stamened.
pleiomerous (plī-om'-er-us)
Pleionema* (plī-on-ē'-mạ)
Pleiospilos* (plī-ŏ-spī'-los)
pleiotropy (plī-ot'-rŏ-pi)
pleiotropic (plī-ŏ-trop'-ik)
Pleistocene (plīs'-tŏ-sēn)
pleocleis (plē'-ŏ-klīs)
Pleocnemia (plē-ok-nē'-mi-ạ)
Pleodorina (plē-ŏ-dō-rī'-nạ)
pleogamy (plē-og'-am-i)
Pleomele* (ple-om'-el-ē)
plerome (plē'-rōm)
plerosis (plê-rō'-sis)
Plesiochelys (plē-si-ok'-e-lis)
Plesiops (plē'-si-ops)
Plesiosaurus (plē-si-ŏ-sôr'-us)
Plesippus (plē-sip'-us)
Plethodon (pleth'-ŏ-don)
Plethopsis (pleth-op'-sis)
Pleurodelidae (plū-rŏ-del'-i-dē)
Pleurodira (plū-rŏ-dī'-rạ)
Pleuronichthys (plū-rŏ-nik'-this)
plexus (pleks'-us, pl. pleks'-us; also, plex'-us-ez)
plica (plī'-kạ)
plicate (plī'-kāt)
plicature (plik'-à-tûr)
plicatus (plik-ā'-tus) folded.
pliciform (plis'-i-fôrm)
Pliocercus (plī-ŏ-serk'-us)
Pliohippus (plī-ŏ-hip'-us)

Ploceidae (plō-sē'-i-dē)
Plocepasser (plō-sē-pas'-êr)
Ploceus (plō'-sê-us)
Plocama* (plok'-am-ạ)
Plodia (plō'-di-ạ)
Ploiariidae (plō-i-ar-ī'-i-dē)
Ploima (plō'-i-mạ)
plover (pluv'êr)
Pluchea* (plū'-ke-ạ)
Plumbago* (plum-bā'-gō)
Plusiidae (plū-sī'-i-dē)
Plutellidae (plū-tel'-i-dē)
Pluvialis (plū-vi-ā'-lis)
Poa* (pō'-ạ)
Podabrus (pod-ab'-rus)
podagricus (pod-ag'-ri-kus) gouty.
podarthrum (pō-dâr'-thrum)
podeon (pōd'-ê-ōn)
podetium (pȯ-dē-shi-um)
podex (pō-deks)
Podica (pod'-i-kạ)
podical (pod'-ik-al)
Podiceps (pod'-i-seps)
podilegous (pō-di-lē'-gus)
podilegus (pod-i-lē'-gus)
podium (pō'-di-um)
Podocarpus* (pod-ȯ-kâr'-pus)
podocephalous (pod'-ȯs-ef'-al-us)
Podoces (pȯ-dō'-sēz)
Podogymnura (pod-ȯ-jim'-nŭ-rạ)
Podolepis* (pod-ol'-ep-is)
podomere (pod'-ȯ-mēr)

Podophrya (pod-ŏ-frī'-ạ)
Podophyllum* (pod-ŏ-fil'-um)
Podostemon* (pod-ŏ-stē'-mon)
podotheca (pod-ŏ-thē'-kạ)
Podura (pō-dū'-rạ)
Poduridae (pō-dū'-ri-dē)
podzol (pod'-zol)
Poeocetes(po-ē-sē'-tēz)
Poecilichthys (pē-sil-ik'-thiz)
poecilogony (pē-si-log'-ŏ-ni)
Poephagus (pō-ef'-ạ-gus)
Pogogyne* (pō-goj'-in-ē)
Pogonia* (pō-gō'-ni-ạ)
pogonion (pō-gō'-ni-on)
Pogonomyrmex (pō-gōn-ŏ-mir'-mex)
Pogostemon* (pō-gŏ-stē'-mōn)
Poicephalus (poy-sef'-al-us)
Poinciana* (poyn-si-ā'-nạ)
Polemonium* (pol-e-mō'-ni-um)
Polianthes* (pol-i-an'-thēz)
polifolius (pol-i-fol'-i-us, pol-i-fō'-li-us) with leaves
    like *Germander*, *Teucrium polium.*
Polinices (pol-i-nī'-sēz)
Poliodon (pol-i'-ŏ-dōn)
Polioptila (pol-i-op'-ti-lạ)
Polistidae (pŏ-lis'-ti-dē)
Polistes (pŏ-lis'-tēz)
politus (pol-ī'-tus) polished.
Polium* (pol'-i-um)
pollen (pol'-en)
polster (pol'-stêr)
Polyborus (pol-ib'-ŏ-rus)

Polycaon (pol-i-kā'-on)
Polycarpon* (pol-i-kâr'-pon)
Polycentropus (pol-i-sen'-trŏ-pus)
Polycera (pol-is'-e-rą)
Polychrosis (pol-ik-rō'-sis)
Polychrus (pol'-i-krus)
Polyctenidae (pol-i-ten'-ĭ-dē)
polyembryony (pol-i-em'-bri-ŏ-ni)
Polygala* (pol-ig'-à-lą)
Polygnotus (pol-ig-nō'-tus)
Polygonatum* (pol-ig-on-āt'-um)
Polygonella* (pol-ig-on-el'-ą)
Polygonum* (pol-ig'-on-um)
polyhybrids (pol-i-hī'-bridz)
Polymitarcidae (pol-i-mi-târ'-si-dē)
Polynices (pol-i-nī'-sēz)
Polyphaga (pol-if'-ag-ą)
polyphemus (pol-i-fē'-mus) many-voiced.
Polyplacophora (pol-i-pla-kof'-ŏ-rą)
polyploidy (pol-i-ploy'-di)
Polypodium* (pol'-i-pō'-di-um, pol-ip-od'-i-um)
Polypogon* (pol-i-pō'-gōn)
Polypremum* (pol-ip'-rem-um)
polyrhizus (pol-i-rī'-zus) many-rooted.
Polyscias* (pol-is'-si-as)
Polystichum* (pol-is'-tik-um)
Polystoechotidae (pol-i-stē-kot'-i-dē)
Polytaenia* (pol-i-tē'-ni-ą)
polytrichous (pol-it'-rik-us)
polytrophic (pol-i-trof'-ik)
Pomaderris* (pō-ma-der'-is)
pomarine (pom'-à-rīn, pom'-à-rin)

Pomatias (pō-mā′-ti-as)
pome (pōm)
pomegranate (pom-gran′-ăt, pum′-gran-ăt)
Pomoxis (pŏ-moks′-is)
Ponera (pon-ē′-ra̧, pŏ-nē′-ra̧)
Poneneridae (pon-er′-i-dē)
Ponicrus* (pon-ik′-rus)
Ponjidae (pon′-ji-dē)
ponogen (pon′-ŏ-jen)
Pontederia* (pon-tē-dē′-ri-a̧)
Pontia (pon′-ti-a̧)
Popillia (pop-il′-i-a̧)
poplar (pop′-lâr)
popliteal (pop-lit′-ê-al, pop-li-tē′-al)
Populus* (pō′-pul-us)
Porana* (pôr-ā′-na̧)
Porcellana (pôr-se-lā′-na̧)
poricidal (pō-ri-sī′-dal)
Porites (pŏ-rī′-tēz)
Porphyrocoma* (pôr-fir-ok′-om-a̧)
porrectus (pôr-ekt′-us)
porrifolius (por-i-fol′-i-us, por-i-fō′-li-us) with
leaves like leek.
Porthetria (pôr-thē′-tri-a̧)
Portulaca* (pôr-tu-lā′-ka, por-tu-la′-ka̧)
Porzana (pôr-zā′-na̧)
posterior (pos-tē′-ri-êr)
posthumous (pos′-tû-mus)
Potamanthidae (pot-am-an′-thi-dē)
Potamochoerus (pot-à-mŏ-kē′-rus)
Potamogale (pot-à-mog′-à-lē)
Potamogeton* (pot-à-mŏ-jē′-ton)

Potamophis (pot-am'-of-is)
potency (pō'-ten-si)
Potentilla* (pō-ten-til'-ạ)
Poterium* (pot-ē'-ri-um)
Potoos (po'-to͞os)
Potorous (pot-o̊-rō'-us)
potto (pot'-ō)
praecox (prē'-koks) before time, immature.
Prasanthea* (pras-an'-the-ạ)
pratensis (prā-ten'-sis) growing in meadows.
Pratincoles (prå-tin-kōl'-ēz)
pratincolus (prå-tin-kōl'-us) meadow inhabiting.
predator (pred'-å-tôr)
Prenanthes* (prē-nanth'-ēz)
prenanthoides (prē-nanth-o-ī'-dēz) with drooping
    leaves or flowers.
preparator (prē-par'-å-tôr, prep-ar'-å-têr)
Presbytes (pres-bī'-tēz)
pretiosus (pret-i-ō'-sus) valuable, at much ex-
    pense.
primaevus (prī-mē'-vus) young.
primordial (prī-môr'-di-al)
Primula* (prim'-ŭ-lạ, prī'-mŭ-lạ)
primigenius (prī-mi-je'-ni-us) first formed.
primiveris (prī-mi-vē'-ris) first of spring.
princeps (prin'-seps) first, in front, most eminent.
Prinia (prin'-i-ạ)
Priodontes (prī-o̊-don'-tēz)
priscus (pris'-kus) first, primitive, of olden times.
pristine (pris'-tin, pris'-tīn)
proboscideus (prō-bo-sid'-e-us) with similar nose.
Proboscidia (prō-bo-sid'-i-ạ)

proboscis (prŏ-bos′-sis, pl. prŏ-bos′-i-dēz)

Procavia (prŏ-kā′-vi-ạ)

procerus (prō-sē′-rus) tall, long, large, extended.

procerus (prō′-ser-us) a muscle of the nose.

Procinura (prō-sin-ūr′-ạ)

Procnias (prok′-ni-as)

proctodeum (prok-tŏ-dē′-um)

procumbens (prō-kum′-benz) bending down, lying along the ground.

Procyon (prō′-si-on)

Prodenia (prō-dēn′-i-ạ)

Prodidomus (prŏ-did′-ŏ-mus)

Prodoxus (prō-doks′-us)

prodromus (prod′-rŏ-mus)

Proechimys (prō-ēk-ī′-mis, prō-ek′-i-mis.)

progamic (prō-gam′-ik)

prognathus (prog′-nath-us)

Progne (prog′-nē)

Proiphys* (prō′-if-is)

proliferate (prō-lif′-êr-āt)

proliferus (prō-li′-fêr-us) bearing progeny, reproducing freely.

prolix (prō′-liks) extended, long.

prolixus (prō-liks′-us) stretched out, long; also, broad.

Promerops (prom′-e-rops)

prophylactic (prō-fi-lak′-tik)

propinquity (prō-pin′-kwi-ti)

propodium (prō-pō′-di-um)

prorsal (prôr′-sal)

prosenchyma (pros-eng′-ki-mạ)

Prosopidiae (prō-sŏ-pid′-i-dē)

Prosopis* (pros-ō′-pis, prŏ-sō′-pis)
Prosthocereus (pros-thŏ-sē′-re-us)
protandry (prŏ-tan′-dri)
protegulum (prō-teg′-ú-lum)
Proteidae (prō-tē′-i-dē)
Proteides (prō-tē′-id-ēz)
protein (prō′-tė-in)

Prosopis <Gr. *prosŏpis*, a kind of plant. The first *o* is short. Pronounced: pros-ō′-pis; also, prŏ-sŏ′-pis.

Proteles (prot′-e-lēz)
Protentomidae (prō-ten-tom′-i-dē)
Proterospongia (prō-te-rŏ-spun′-ji-ạ)
proterothesis (prō-te-rŏ-thē′-sis)
Proterozoic (prō-te-rō-zō′-ik)
proteus (prō′-te-us)
prothorax (prō-thō′-raks)
Protonotaria (prō-tŏ-nō-tā′-ri-ạ)
Protoparce (prō-tŏ-par′-sē)
Protophyta (prō-tof′-it-ạ)
protopodite (prō-tō′-pŏ-dīt, prō-top′-ŏ-dit)
Protopterus (prō-top′-te-rus)

Protura (prō-tū′-rạ)
provectus (prō-vek′-tus) advancing, increasing.
pruinosus (prū-i-nō′-sus) full of hoarfrost.
Prumnopitys* (prum-nop′-it-is)
Prunella* (prū-nel′-ạ)
psalterium (sôl-tē′-ri-um)
Psamma* (sam′-ạ)
Psammocharidae (sam-ô-kar′-i-dē)
Psathyrotes* (sath-i-rō′-tēz)
Pselaphidae (sĕ-laf′-i-dē)
Pselaphus (sel′-ạ-fus)
Psephenus (sef-ēn′-us)
Psephotus (sef-ōt′-us)
Pseudacris (sūd-a′-kris)
Pseudechis (sūd′-ek-is)
Pseudemys (sū′-de-mis)
Pseudochirus (sū-dô-kīr′-us)
Pseudois (sū′-dô-is)
Pseudolarix* (sū-dô-la′-riks)
Pseudomethoca (sū-dô-meth′-ok-ạ)
Pseudoplisus (sū-dop-lī′-sus)
pseudopodium (sū-dô-pō′-di-um)
Pseudotsuga* (sū-dô-tsū′-gạ)
Psidium* (sid′-i-um, psid′-i-um)
Psilactis* (sī-lak′-tis)
Psilonema (sī-lon-ē′-ma)
Psilonotus (sī-lô-nō′-tus)
Psilophyta (sī-lof′-it-ạ)
Psilophyton (sī-lof′-i-ton)
Psilostrophe* (sī-lō′-stro-fē)
Psilotum* (sī-lō′-tum)
Psithyrus* (sith′-i-rus)

psittaceus (sit-ā'-se-us) parrot-like.
Psittacus (sit'-à-kus)
psoas (sō'-as, psō'-as)
Psocidae (sos'-i-dē)
Psocinella (sō-sin-el'-ạ)
Psocoptera (sō-kop'têr-ạ)
Psocus (sō'-kus)
Psolidae (pso'-li-dē)
Psolus (psō'-lus)
Psophocarpus* (sō-fô-kâr'-pus)
psora (sō'-rạ) the itch.
Psoralea* (psō-ral'-e-ạ)
Psoroptes (sō-rop'-tēz)
Psychidae (sī'-ki-dē)
Psychodidae (sī-kōd'-i-dē)
Psychomyiidae (sī-kô-mī'-i-dē)
Psychotrophum (sī-kot'-rof-um)
psychrometer (sī-krom'-ê-têr)
Psydrax (sid'-raks)
Psylla (si'-lạ)
Psythirus (psith'-i-rus)
ptarmigan (târ'-mi-gan)
Ptelea* (tel'-e-ạ, tē'-lê-ạ)
Pteranodon (ter-an'-ô-don)
pteridophyte (ter-i'-do-fīt, ter'-i-do-fīt)
Pteridophyta* (ter-i-dof'-i-tạ)
Pteris* (ter'-is, pter'-is)
Pterocles (ter'-ô-klēz)
Pterocletes (ter-ô-klē'-tēz
Pterodactyl (ter-ô-dak'-til)
Pterodroma (ter-od'-rô-mạ)
Pteromalidae (ter-ô-mal'-i-dē)

PTEROMYS

Pteromys (ter'-ŏ-mis)
Pteronarcidae (ter-ŏ-när'-si-dē)
Pterophora (ter-of'-ôr-ạ)
Pterophoridae (ter-ŏ-fôr'-i-dē)
Pterophorus (ter-of'-ôr-us)
Pteropus (ter'-ŏ-pus)
Pterospora* (ter-os'-pŏ-rạ)
Pterostichus (ter-os'-tik-us)
pterotus (ter-ō'-tus) winged, with handles.
pterygius (ter-i'-ji-us) winged, with wing-like spot.
Pterygota (ter-i-gō'-tạ)
Ptilichthys (til-ik'-this)
Ptilimnium* (til-im'-ni-um)
ptilinum (til'-i-num)
Ptilocnema* (til-ok-nē'-mạ)
Ptilodexia (til-ŏ-deks'-i-ạ)
Ptilomeris* (til-om'-er-is)
Ptiloris (til-ôr'-is)
ptilosis (til-ō'-sis, ptil-ō'-sis)
Ptilostephium* (til-os-teph'-i-um)
Ptilota* (til-ō'-tạ, ti-lō'-tạ)
Ptinidae (tin'-i-dē)
Ptinobius (tin-ob'-i-us)
Ptinus (tī'-nus)
ptomain (ptō'-mā-in, ptō'-mān, tō'-man)
Ptyas (tī'-as)
Ptychoramphus (tī-kŏ-ram'-fus, tik-or-am'-fus)
Ptychosperma* (tī-kŏ-spêr'-mạ, tik-os-pêr'-mạ)
Ptycozoon (tī-kŏ-zō'-on)
Ptylichthys (tī-lik'-this)
ptyocrinus (tī-ok'-rin-us)
ptyxis (tik'-sis)

puberulent (pū-ber′-ŭl-ent)

pubescens (pū-be′-senz) downy, slightly hairy.

Pueraria* (pū-er-ā′-ri-ạ)

pulchellus (pul-kel′-us) somewhat beautiful.

pulcher (pul′-ker) handsome, beautiful, excellent.

pulegium (pū-le′-ji-um) pennyroyal.

Pulicaria* (pū′-li-kā′-ri-ạ)

pulicarius (pū-li-kā′-ri-us) of or belonging to fleas.

Pulicidae (pū-lis′-i-dē)

pullus (pu′-lus) dark-colored, dusky.

pulsellum (pul-sel′um)

Pultenaea* (pul-ten-ē′-ạ)

pulverulent (pul-ver′-ŭ-lent)

pulverulentus (pul-ver-u-len′-tus) dusty.

pulvillus (pul-vil′-us, pl. pul-vil′-ī)

pulvinus (pul-vī′-nus, pl. pul-vī-nī)

pumilus (pū′-mi-lus) dwarfish.

punctate (pung′-ktāt)

punctatus (pung-ktā′-tus) marked with dots.

Punctum (pung′-ktum)

pungens (pun′-jenz) piercing.

Punica* (pū′-nik-ạ)

puniceus (pū-ni′-se-us) reddish, red, purple.

Pupipara (pū-pi′-pa-rạ)

purpuraceus (pûr-pûr-ā′-se-us)

purpureus (pûr-pū′-re-us) purple; also, red, reddish, brilliant.

purus (pū′-rus) clean, pure; also, unadorned, free
from spots, clear, bright.

pusillus (pus-il′-us) small, insignificant.

putorius (pū-to′-ri-us) with foul odor, rottenness.

putus (put′-us) pure, clear, unmixed.

Puya* (pū'-yạ)
Pycnanthemum* (pik-nan'-the-mum)
Pycnogonum (pik-nog'-ŏ-num)
Pycnonotidae (pik-nŏ-not'-id-ē)
Pycnonotus (pik-nŏ-nōt'-us)
Pygaera* (pī-gē'-rạ)
pygal (pī'-gal)
Pygathrix (pī'-gȧ-thriks)
Pygidicranidae (pī-jid-i-kran'-i-dē)
pygidium (pī-jid'-i-um)
Pygopodes (pī-gop'-ŏ-dēz)
Pygopus (pī'-gŏ-pus)
Pygoscelis (pī-gos'-e-lis)
pygostyle (pī'-gŏ-stīl)
pylangium (pi-lan'-ji-um, pī-lan'-ji-um)
pylic (pī'-lik)
pyloris (pī-lō'-ris)
Pyracantha* (pir-ak-anth'-ạ)
Pyragra (pir-a'-grạ)
Pyralidae (pi-ral'-i-de)
Pyralis* (pir'-ạl-is)
pyramidal (pir-am'-id-al)
Pyrausta (pī-rô'-stạ)
Pyraustidae (pī-rô'-sti-dē)
pyrene (pī'-rēn)
pyrenocarp (pī-rē'-nŏ-kârp)
pyrenoid (pī-rē'-noid)
Pyrethrum* (pī-reth'-rum, pir'-eth-rum, pir-ē'-
    thrum)
pyriform (pir'-i-fôrm)
Pyrochroa (pī-rok'-rŏ-ạ)
Pyrola* (pir'-ŏ-lạ)

Pyrophila* (pī-rof'-i-lạ)
Pyrophorus (pī-rof'-ô-rus)
Pyrostegia* (pī-rô-stē'-ji-ạ)
Pyrrhocorax (pi-rô-kôr'-aks, pi-rok'-ô-raks)
Pyrrhopappus* (pi-rô-pap'-us, pir-ô-pap'-us)
Pyrrhuloxia (pir-ōo-lok'-si-ạ)
Pyrrosia* (pir-rō'-si-ạ)
Pyrularia* (pir-ŭ-lā'-ri-ạ)
Pyrus* (pir'-us)
Pythium* (pith'-i-um)
Python (pī'-thon)
Pythonium* (pī-thō'-ni-um)
Pyticera (pit-is'-er-ạ)
Pyxidanthera* (piks-id-an-thē'-rạ, piks-id-anth'-
    er-ạ)
pyxis (pik'-sis)

Q

quadrangulus (kwod-ran'-gu-lus) four-cornered.
quadrifidus (kwad-rif'-id-us) divided in four.
Quadrumana (kwod-rū'-man-ạ)
quadrupedal (kwod-rōo'-pe-dal)
Qualea* (kwā'-le-ạ)
Quamoclit* (kwa-mok'-lit)
quarantine (kwôr'-an-tēn)
quartile (kwôr'-til)
Quassia* (kwäsh'-i-ạ)
Quelea (kwē'-lê-ạ)
querceticola (kwer-se-tik'-ol-ạ) oak dwelling.
quercetum (kwer-sē'-tum) an oak-wood.
Quercus* (kwer'-kus)
Querquedula (kwer-kwed'-ŭ-lạ)

querulus (kwer'-ru-lus) plaintive.
Quincula (kwin'-ku-lạ)
quincunx (kwin'-kungks)
quinquemaculatus (kwin-kwe-mak-ul-ā'-tus) five spotted.
quintuple (kwin'-tu-pl)
quintuplets (kwin'-tu-plets)
Quiscalus (kwis'-kặ-lus)
Quisqualis* (kwis-kwā'-lis)

R

rabies (rab'-i-ēz, rā'-bi-ēz)
raceme (rā-sēm')
racemosus (rā-sẽ-mō'-sus) full of clusters, clustered.
rachial (rā'-ki-al)
rachilla (ra-ki'-lạ)
rachiodont (rāk'-i-ộ-dont)
rachis (rā'-kis, pl. rā'-ki-dēz)
racial (rā'-shal)
radicans (rā-dī'-kanz) taking root.
radicant (rad'-i-kant)
radicatus (rā-dī-kā'-tus) rooted.
radicivorous (rad-is-iv'-ôr-us)
radicose (rad'-i-kōs)
Radiola* (rad'-i-o-lạ)
Radiolaria (rad-i-o-lā'-ri-ạ, rād-i-o-lā'-ri-ạ)
radiosus (rad-i-ō'-sus) radiant, giving forth many beams.
radius (rā'-di-us)
radix (rā'-diks, pl. rā'-di-sēz)
radula (rad'-ŭl-ạ)

Rallus (ral'-us)
ramentactaceus (rā-men-tā'-se-us)
ramigerous (ram-ij'-êr-us)
ramose (rā'-mos, rȧ-mōs')
ramosus (rā-mō'-sus) with many branches, branching.
ramulosus (rā-mŭ-lō'-sus) full of branches or twigs.
ramus (rā'-mus, pl. rā'-mī)
Rana (rā'-nạ)

Rana <L. *rana*, a frog. Pronounced: rā'-na, not ra'-nạ.

Ranatra (ran'-at-rạ)
Rangifer (ran'-ji-fêr)
Raniceps (ran'-i-seps)
Ranidae (ran'-i-dē)
raniform (rā'-ni-fôrm, ran'-i-fôrm) frog-shaped.
Ranunculaceae* (rā-nung-kul-ā'-sê̇-ē)
ranunculoides (rā-nung-kul-o-ī'-dēz)
Ranunculus* (rā-nung'-kul-us)
Raoulia* (rȧ-ōō'-li-ạ)
Rapa* (rā'-pạ)
Raphanus* (raf'-ȧ-nus)
raphe (rā'-fē, pl. rā'-fī)

Raphia* (rā'-fi-ạ, raf'-i-ạ)
raphid (raf'-id, pl. raf'-id-ēz)
Raphidiidae (raf-id-ī'-i-dē)
Raphidophyllum* (raf-i-dǒ-fil'-um)
raphidus (raf'-i-dus)
Raphiolepis* (raf-i-ol'-ep-is)
Raphistemma (raf-is-tem'-ạ)
Raphus (raf'-us)
rapunculoides (rā-pung-ku-lo-ī'-dēz) like a little turnip.
Rapunculus* (rā-pung'-kul-us, rap-ung'-kŭ-lus)
rariflorus (rā-ri-flō'-rus) not dense-flowered.
rarus (rā'-rus) thin, dispersed.
rasorial (rȧ-sō'-ri-al)
Ratibida* (rat-ib'-id-ạ)
ratio (rā'-shō)
Ratitae (rat'-ī-tē)
ratite (rat'-īt)
Ratufa (rat-ū'-fạ)
ravidus (rā'-vi-dus) grayish, dark-colored.
ravus (rā'-vus) grayish-yellow, gray.
reclinatus (rek-lin-ā'-tus) turned or bent downward, bent back.
rectrices (rek-trī'-sēz, sing. rek'-triks)
rectrix (rek'-trix, pl. rek-trī'-sēz)
Recurvirostra (rê-kûr-vi-ros'-trạ)
redimiculum (red-i-mī'-ku-lum) a band, a headband.
Redunca (rê-dung'-kạ)
reduncus (rê-dung'-kus) curved or bent back.
Reduviidae (red-ŭ-vī'-i-dē)
Reduviolus (red-ŭ-vī'-ol-us)

reflexus (rē-flex'-us) bending back.

regalis (reg-ā'-lis) kingly, royal.

regius (rē'-ji-us) royal.

Regulus (reg'-u-lus)

Reithrodontomys (rī-thrŏ-don'-tŏ-mis)

relict (rel'-ikt)

remex (rē'-meks, pl. rem'-i-jēz)

remiges (rem'-i-jēz, sing. rē'-meks)

remigrant (rem'-ig-rant)

Remora (rem'-ôr-ạ)

remotus (rem-ō'-tus) distant, remote.

ren (ren, pl. rē'-nēz)

renal (rē'-nal)

Renanthera* (rē-nan-thē'-rạ)

renascent (rē-nas'-ent)

reniform (ren'-i-fôrm, rē'-ni-form)

Renilla (ren-il'-ạ)

repand (rē-pand')

reparative (rē-par'-ȧt-iv)

repellant (rē-pel'-ant)

repens (re'-penz) unexpected, unlooked for, sudden.

repletes (rē-plēts')

replicatile (re-plik'-ȧ-til, rep-lik'-ȧ-tīl)

replum (rep'-lum) a door-case.

reptans (rep'-tanz) creeping.

reptile (rep'-til)

resartus (res-âr'-tus) restored, patched.

research (rē-sêrch')

resectus (res-ek'-tus) cut off.

Reseda* (rē-sē'-dạ, res'-ē-dạ)

reservoir (res'-êr-vwôr, rez'-êr-wvâr)

resolutus (res-ol-ū'-tus) released, loosened.
respiratory (rê-spīr'-ā-tō-ri, res'-pi-rȧ-tō-ri)
reticulatus (rē-ti-ku-lā'-tus) net-like.
retifer (rē'-ti-fer) net-bearing.
Retinospora (rē-tī-nos'-pôr-ạ)
retrograde (ret'-rô-grād)
retromorphosis (ret-rô-môr-fō'-sis)
retrusion (rê-trū'-zhun)
retrorse (rē-trôrs')
retrostalis (ret-rô-stal'-sis)
retrusus (ret-rū'-sus) distant, hidden.
retusus (ret-ū'-sus) blunted, dull.
Reynosia* (rā-nō'-shi-ạ)
Rhabdocoelida (rab-dô-sē'-li-dạ)
Rhachianectes (rā-ki-an'-ek-tēz)
Rhacoma* (rak-ō'-mạ)
Rhacomitrium* (rak-ô-mit'-ri-um)
Rhacophorus (rȧ-kof'-ô-rus)
Rhagadiolus* (rā-gȧ-dī'-ol-us)
Rhagionidae (rā-ji-on'-i-dē)
Rhagodia* (rȧ-gō'-di-ạ)
Rhagoletis (rȧ-gô-lē'-tis)
rhagon (rag'-on, rā'-gon)
Rhamnidium* (ram-ni'-di-um)
rhamnifolius (ram-ni-fol'-i-us, ram-ni-fō'-li-us) with Rhamnus-like leaves.
Rhamnus* (ram'-nus)
Rhamphastos (ram-fast'-os)
Rhampholeon (ram-fō'-lê-on, ram-fō'-le-ōn)
Rhamphorhynchus (ram-fô-ring'-kus)
Rhanis (ran'-is)
Rhaphanistrum* (raf-an-is'-trum)

rhaphe (rā'-fē)
Rhaphidophora* (raf-id-of'-ôr-ạ)
Rhapis* (rā'-pis)
Rhaponticum* (rā-pon'-tik-um)
Rheomys (rē'-ŏ-mis)
Rheumaptera (rū-map'-têr-ạ)
Rhexia* (rek'-si-ạ)
Rhinanthus* (rī-nan'-thus)
Rhineura (rī-nū'-rạ)
Rhinocerus (rī-nos'-er-us)
Rhinocheilus (rī-nŏ-kī'-lus)
Rhinotermitidae (rī-not-êr-mit'-i-dē)
Rhinotora (rī-not'-ôr-ạ)
Rhipiphorus (rip-if'-ôr-us)
Rhipsalis* (rip'-sal-is)
Rhiptoglossa (rip-tŏ-glos'-ạ, rip-tŏ-glō'-sa)
rhizanthous (rī-zan'-thus)
rhizautoicus (rī-zô'-toy-kus)
Rhizina (rī-zī'-nạ)
Rhizomys (rī'-zŏ-mis)
Rhizophora (rī-zof'-ŏ-rạ)
Rhodea* (rō'-dê-ạ)
Rhodiola* (rod'-i-ōl-ạ, rō-di-ōl'-ạ, rô-dī'-ol-ạ)
Rhododendron* (rod-ôd-en'-dron, rō-dŏ-den'-dron)
Rhodomela* (rō-dom'-e-lạ)
Rhodope (rō'-dŏ-pē)
Rhodophycophyta (rod-ŏ-fī-kof'-it-ạ, rō-dŏ-fī-kof'-
    it-ạ)
Rhodostethia (rod-ŏ-stē'-thi-ạ)
Rhodothamnus* (rod-ôth-am'-nus, rō-dŏ-tham'-
    nus)
Rhodotypos* (rod-ot'-ip-os)

Rhomboplites (rom-bop-lī'-tēz)
Rhopalocera (rō-pal-os'-er-ạ)
Rhopalomera (rō-pal-om-ē'-rạ)
Rhus* (rus, rūs)
Rhyacophilidae (rī-a-kô-fil'-i-dē)
Rhyacotriton (rī-ak-ô-trī'-ton)
Rhymbus (rim'-bus)
Rhynchetus (ring-kē'-tus)
Rhynchobdelida (ring-kob-del'-id-ạ)
Rhynchocyon (ring-kos'-i-on)
Rhynchophanes (ring-kof'-ậ-nēz)
Rhynchops (ring'-kops)
Rhynchosia* (ring-kō'-shi-ạ)
Rhynchospora* (ring-kos'-pō-rạ)
Rhynchostoma (ring-kos'-tô-mạ)
Rhynchotragus (ring-kô-trāg'-us)
Rhynchotus (ring-kō'-tus)
Rhyncophora (ring-kof'-ôr-ạ)
Rhyssa (ris'-ạ)
Rhyssodes (ri-sō'-dēz)
rhytidome (rit'-i-dōm)
rhytidophyllum (rit-id-ô-fil'-um)
Rhytina (ri-tī'-nạ)
Ribes* (rī'-bēz)
Ricinidae (ris-in'-i-dē)
Ricinulei (ri-sin-ū'-lê-ī)
Ricinus* (ris'-in-us)
Ricotia* (rī-kō'-ti-ạ)
rigidulus (rij-id'-u-lus) rigid, stiff, hard.
rigidus (rij'-ji-dus) stiff, hard, not flexible.
Rigiopappus* (rij-i-ô-pap'-us)
Rima (rī'-mạ)

rimosus (rī-mō′-sus) full of cracks or fissures.

ringens (rin′-jenz) grinning, snarling.

Riparia (rip-ār′-i-ạ)

riparius (rip-ā′-ri-us) frequenting river banks.

risorius (rī-sôr′-i-us)

Rissa (ris′-ạ)

rivalis (rī-vā′-lis) of or belonging to a brook.

rixosus (riks-ō′-sus) quarrelsome.

robiginosus (rō-bī-ji-nō′-sus) rusty.

Robinia* (rob-in′-i-ạ, rô-bin′-i-ạ)

Rodentia (rō-den′-shi-ạ)

Romalea (rō-mā′-lê-ạ)

Romneya* (rom′-ne-ạ, rom-nē′-yạ)

Rondeletia* (ron-del-ē′-ti-ạ)

root (rōōt)

Rorippa* (rō-rip′-ạ)

Rosa (ros′-ạ, rō′-zạ, rō′-sạ)

roseus (ros′-se-us) rose-colored.

rosmarinifolius (ros-ma-rī-ni-fol′-i-us, ros-ma-rī-ni-fō′-li-us) with leaves like rosemary.

Rosa <L. *rosa*, the rose <Gr. *rhodon*. Although the English word "rose," and the girl's name "Rose" are pronounced with a long *o*, the Latin rosa, has the *o* short. Pronounced: ro′-sa. Because of long usage rō-za is considered acceptable.

rostralis (ros-trā′-lis) of or concerning a beak or snout.

rostratus (ros-trā′-tus) beaked.

rosulatus (ros-u-lāt′-us) resembling a rose, arranged in a rosette.

Rotala* (rot-ā′-lạ)

rotatus (rot-ā′-tus) a turning round.

rotifer (rō′-ti-fêr)

Rotifera (rō-ti′-fêr-ạ)

rotula (rot′-ū-lạ)

rotundus (rot-un′-dus) round, wheel-shaped.

rouleaux (roo-lō′)

rubellus (rub-el′-us) a little bit red.

rubens (rub′-enz) becoming red.

ruber (rub′-er) red.

rubeta (rub-ē′-tạ) a kind of poisonous toad.

rubeus (rub′-e-us) red, reddish.

Rubia* (rub′-i-ạ)

Rubicola (rub-ik′-ô-lạ)

rubinus (rub′-in-us) red.

Rubus* (rub′-us)

Rudbeckia* (rud-bek′-i-ạ)

ruderalis (rud-er-āl′-is) growing in waste places or among rubbish.

rudis (rud′-is) rough, raw, untilled.

Ruellia* (rū-el′-i-ạ)

rufescens (rū-fes′-senz) becoming red, reddish.

rufidus (rū′-fi-dus) somewhat red.

rufinism (rū′-fin-izm)

rufus (rū′-fus) red, reddish; also, red-headed.

ruga (rū′-gạ) a crease or wrinkle.

rugilobus (rū-jil′-ob-us) with wrinkled lobes.

ruginosus (rū-jin-ō′-sus) wrinkled.

rugosus (rū-gō′-sus) wrinkled, corrugated, shriveled.

ruidus (ru′-i-dus) rough.

Rumex* (roo‾′-meks)

runcinatus (run-sin-ā′-tus) planed off, made smooth.

rupester (roo‾-pes′-têr) growing on rocks.

Rupicola, Cock of the Rock <L. *rūpes*, genit. *rupis*, a rock+*colō*, to inhabit.
Pronounced: roo‾-pik′ŏ-lạ (accent on the antepenult), not rū-pi-kŏ′-lạ.

rupestrine (roo‾-pes′-trin)

Rupicapra (roo‾-pi-kā′-prạ, roo‾′-pi-kap′-rạ)

Rupicola (roo‾-pik′-ŏ-la)

rupicolous (roo‾-pik′-ŏ-lus)

rupicolus (roo‾-pik′-ŏ-lus) rock-dwelling.

Rusa (roo‾′-sạ)

Ruscus* (rus′-kus)

rusticus (rus′-ti-kus) rustic, rural.

Ruta* (roo‾′-tạ)

Rutelidae (rŭ-tel′-i-dē)

rutilus (rut′-il-us) red, ruddy.

Rynchophanes (ring-kof′-ån-ēz)
Rytiginia* (rit-ij-in′-i-ạ)

S

Sabal* (sā′-bal)
Sabbatia* (sab-ā′-ti-ạ)
sabine (sā′-bīn)
sabulus (sab′-ū-lus)
saccharatus (sak-å-rā′-tus) sugary, sweet.
saccharine (sak′-å-rin, sak′-a-rīn)
Saccharum* (sak′-å-rum)
sacciferous (sak-sif′-er-us)
sacciform (sak′-si-fôrm)
Sacciolepis* (sak-i-ol′-ep-is)
Saccochilus* (sak-ok-ī′-lus)
Saccolabium* (sak-ol-ab′-i-um)
Saccophora (sak-of′-ô-rạ)
Sacculina (sak-ū-lī′-nạ)
sacer (sas′-er) holy, sacred.
Sacodes (så-kō′-dēz)
sacrarium (så-krâr′-i-um)
Sagartia (så-gâr′-ti-ạ)
Sageretia* (sag-er-ē′-shi-ạ, sag-er-ē′-ti-ạ)
Sagina* (saj-ī′-nạ)
sagitta (saj-it′-ạ)
sagittal (saj′-i-tal) pertaining to an arrow.
Sagittaria* (saj-it-tā′-ri-ạ)
sagittifolius   (saj-i-ti-fol′-i-us,   saj-i-ti-fō′-li-us)
  arrow-leaved.
Sagmatias (sag-ma′-ti-us)
Saiga (sī′-gạ, sā′-i-gạ)
Salacia* (sal-ā′-si-ạ)

Salazaria* (sal-a-zâr′-i-ạ)

Saldidae (sal′-di-dē)

salebrosus (sal-e-brō′-sus) rough, uneven, full of bumps.

Salicaria* (sal-ik-ā′-ri-ạ)

salicarius (sal-ik-ār′-i-us) of or pertaining to willows.

salicifolius (sal-is-i-fol′-i-us, sal-is-i-fō′-li-us) willow-leaved.

Salicornia* (sal-i-kôr′-ni-ạ)

salignus (sal-i′-gnus) of willow wood, of willow.

saline (sā′-lin)

Salix* (sal′-iks, sā′-liks)

Salpiglossus* (sal-pi-glos′-us, sal-pi-glō′-sus)

Salpinctes (sal-pingk′-tēz)

salpingectomy (sal-pin-jek′-tô-mi)

Salpinx* (sal′-pingks)

Salsola* (sal′-sô-lạ)

salsuginous (sal-sū′-ji-nus)

saltator (sal-tā′-tôr) a leaper, a dancer.

Salvadora (sal-vȧ-dôr′-ạ)

Salvelinus (sal-ve-lī′-nus)

Salvia* (sal′-vi-ạ)

Salvinia* (sal-vī′-ni-ạ)

Samadera* (sa-mad′-êr-ạ)

Samanea* (sam-ā′-nê-ạ)

Samara* (sam′-ȧr-ạ, sam-ā′-rạ)

samara (sam′-ȧ-rạ, sȧ-mā′-rạ)

Sambucus* (sam-bū′-kus)

Samia (sā′-mi-ạ)

Samolus* (sā′-mol-us, sam′-ol-us)

Sandoricum* (san-dor′-ik-um)

sanguinalis (san-gwi-nā'-lis) bloody, of blood, blood thirsty.

Sanguinaria* (san-gwi-nā'-ri-ą)

saguineus (san-gwi'-ne-us) bloody, of blood, blood-red.

Sanguisorba* (san-gwi-sôr'-bą)

Sanicula* (san-ik'-u-lá)

Sansevieria* (san-se-vēr'-i-ą, san-sev-i-ē'-ri-ą)

Saperda (sap-êr'-dą)

sapidus (sap'-i-dus) good to eat, savory.

sapiens (sap'-i-enz) knowing, of good taste.

Sapindus* (sap-in'-dus, sā-pin'-dus)

Sapium* (sap'-i-um, sā'-pi-um)

Saponaria* (sap-o-nā'-ri-ą)

Sapota* (sa-pō'-tą)

Saprinus (sap'-rin-us)

saprophytic (sap-rŏ-fit'-ik)

Sapygidae (sȧ-pij'-i-dē)

Sarachilus* (sâr-ak-īl'-us)

Sarcina* (sâr'-si-ną)

Sarcobatis* (sâr-kob'-at-is)

Sarcobatus* (sâr-kob'-at-us)

Sarcodina (sâr-kŏ-dī'-ną)

Sarcoglottis* (sâr-kog'-lot-is, sâr-kog-lō'-tis)

Sarcolobus* (sâr-kol'-ŏb-us)

Sarcophagidae (sâr-kŏ-faj'-i-dē)

Sarcophilus (sâr-kof'-il-us)

Sarcopsylla (sâr-kop-sil'-ā)

Sargassum (sâr-gas'-um)

Sargania* (sâr-gan'-i-ą)

Sargus (sâr'-gus)

Sarcobatus <Gr. *sarkos*, flesh+*batos*, a bramble.
Pronounced: sâr-kob'-a-tus, not sar-kō-bā'-tus.

sarmentosus (sâr-men-tō'-sus) twiggy, full of little branches.

Sarothamnus* (sâr-ȯ-tham'-nus)

sarothroides (sar-ōth-ro-ī'-dēz)

sarothrum (sar-ō'-thrum) a broom.

Sarracenia* (sar-as-ē'-ni-ạ)

sartorius (sâr-tō'-ri-us)

Sasa* (sä'-sä)

Sasia (sā'-si-ạ)

sasin (sā'-sin)

Satira (sat-īr'-ạ)

sativus (sat-ī'-vus) planted, that is sown.

saturatus (sat-u-rā'-tus) full of color, rich in color

Satureia* (sat-ů-rē'-i-ạ)

Saturniidae (sat-ûr-nī'-i-dē)

satyr (sat'-êr, sā'-têr)

Sauroglossum* (sȯ-rog-glos'-um, sȯ-rog-glō'-sum)

Sauromalus (sȯ-rȯ-māl'-us)

Sauropsida (sȯ-rop'-sid-ạ)

Saururus* (sȯ-rū'-rus)

saxatilis (saks-ā'-ti-lis) found among rocks.

saxicolous (saks-ik′-ol-us)
Saxifraga* (sak-sif′-rå-gą)
saxosus (saks-ō′-sus) stony, full of rocks.
Sayornis (sā-ôr′-nis)
scaber (skab′-er) rough, scurfy.
Scabiosa* (skab-i-ō′-są, skā-bi-ō′-są)
scabricomus (skab-ri′-ko-mus)
scabrosus (skab-rō′-sus) rough.
scabrous (skab′-rus)

Saxifraga <L. *saxifraga*, "the rock-breaker" <*saxum*, rock+*frangere*, to break. Pronounced: saks-if′-ra-gą, not saks-if-rā′-gą, as we often hear.

Scaevola* (sē′-vol-ą)
Scalaria (skå-lā′-ri-ą)
scalene (skå-lēn′)
scallop (skal′-up, skol′-up)
Scalops (skā′-lops)
Scalopus (skal′-op-us)
Scandix* (skan′-diks)
scansorial (skan-sō′-ri-al)
Scapanus (skap′-à-nus)
scape (skāp)
Scaphiopus (skaf-ī′-ô-pus)
Scaphisoma (skaf-is-ōm′-ą)

Scaphites (skaf-ī'-tēz)
Scapholeberis (ska-fŏ-leb'-êr-is)
Scaphopoda (skaf-op'-ŏ-dą)
scapiodeus (skȧ-poyd'-e-us) scape-like.
Scaptolemus (skap-tŏ-lēm'-us)
scapulare (skap-ŭ-lā'-rē)
Scarabaeidae (skar-ȧ-bē'-i-dē)
scariola (ska-ri-ōl'-ą) wild lettuce.
scarious (skā'-ri-us)
Scatophaga (skat-of'-ag-ą)
Scatophagidae (scat-ŏ-faj'-i-dē)
scaup (skôp)
sceleratus (sel-er-ā'-tus) defiled, vicious, bad.
Scelidosaurus (sel-id-ŏ-sô'-rus)
Scelio (sē'-li-ō)
Scelionidae (sel-i-on'-i-dē)
Sceliphron (sel'-if-ron)
Sceloporus (sê-lop'-o-rus, sel-op'-ôr-us)
Scenedesmus (sē-nê-des'-mus)
Scenopinus (sê-nŏp'-in-us)
Scepsis (skep'-sis)
Schedius (sked'-i-us)
Schedonnardus* (sked-on-âr'-dus)
schidigerus (ski-di'-jer-us) splinter-bearing.
schindylesis (skin-dil-ē'-sis)
schizont (skī'-zont)
Schinus* (skī'-nus)
Schistosoma (skis-tŏ-sō'-mą)
schistosomiasis (skis-tŏ-som-i-ā'-sis)
Schizandra* (skiz-an'-drą)
schizogomy (skiz-og'-a-mi)
Schizoloma* (skiz-ol-ō'-mą)

Schizophragma* (skiz-ôf-rag′-mą)
Schizophyta (skiz-of′-it-ą)
Schizopoda (skiz-op′-ô-dą)
Schizostylis* (skiz-os′-til-us)
Schoenoprasum* (skēn-op′-ras-um)
Schoenus* (skē′-nus)
Sciadocalyx (sī-ad-ok′-al-iks, si-ad-ok′-al-iks)
Sciadopitys* (sī-à-dop′-it-is, si-ad-op′-it-is)
Sciaena (sī′-ē-ną)
Sciagraphia (sī-à-graf′-i-ą)
Sciara (sī′-à-rą)
Scilla* (sil′-ą)
Scincus (sing′-kus)
scion (sī′-on)
Scirpus* (sûr′-pus)
scission (sizh′-un, sish′-un)
Scissirostrum (sis-i-ros′-trum)
scitulus (skit′-u-lus) slender, graceful, elegant.
Sciuropterus (sī-û-rop′-ter-us)
Sciurus (sī-ū′-rus)
Scleranthus* (sklē-ran′-thus)
scleroblast (sklē′-rô-blast)
Sclerochloa* (sklē-rô-klō′-ą)
Sclerodermi (sklē-rō-dêr′-mī)
Sclerolopis* (sklē-rô′-le-pis)
Scleropogon* (sklē-rô-pō′-gōn)
sclerotic (sklê-rot′-ik)
sclerotin (skler′-ô-tin)
Sclerurus (sklē-ur′-us)
scobina (skob-ī′-ną) a rasp.
Scoliidae (skō-lī′-i-dē)
Scoliodon (skô-lī′-ô-don)

Scolopax (skol'-ō̇-paks)
Scolopendrium* (skol-ōp-en'-dri-um)
Scolops (skol'-ops, skō'-lops)
Scolymus* (skol'-im-us)
Scolytidae (skō̇-lit'-i-dē)
Scolytus (skol'-i-tus)
scomberomorous (skom-bē̇-rom'-ō̇-rus)
Scombresox (skom'-bre-soks)
Scoparia* (skō-pā'-ri-ą)
scoparius (skō-pā'-ri-us) a sweeper.
Scopelus (skop'-e-lus)
Scopidae (skop'-i-dē)
scopiform (skō'-pi-fôrm) broom-shaped.
scops (skops)
Scordium* (skôr'-di-um)
scorteus (skôr'-te-us) made of leather.
Scorzonella* (skôr-zon-el'-ą)
Scorzonera* (skôr-zon-ē'-rą)
Scotiaptex (skō̇-ti-ap'-tex, skō-shi-ap'-tex)
Scotophilus (skō̇-tof'-il-us)
Scotornis (skō̇-tôr'-nis)
scrobiculate (skrob-ik'-ū̇-lāt)
scrofa (skrof'-ą) a breeding-sow.
Scrophularia* (skrof-u-lā'-ri-ą)
scrotal (skrō'-tal)
Scrupocellaria (skrū-pō̇-sel-ā'-ri-ą)
scrupulosus (skrū-pul-ō'-sus) rough; also, exactly, carefully.
scurvy (skêr'-vi)
scutatus (skū-tā'-tus) armed with a shield.
Scutellaria* (skū-tel-ā'-ri-ą)

scutellatus (skū-tel-ā′-tus) with shield-like parts, shield-like.

Scutelleridae (skū-tel-er′-i-dē)

scutellum (skūt-el′-um)

Scuticaria (skū-tik-ā′-ri-ạ)

Scutigera (skū-ti′-je-rạ)

Scutula* (skut′-u-lạ)

scutullatus (skut-ul-ā′-tus) diamond-shaped, checkered.

scutum (skū′-tum)

Scydmaenidae (sid-mēn′-i-dē)

Scylla (sil′-ạ)

Scyllarus (sil′-à-rus)

Scymnus (sim′-nus)

Scypha (sī′-fạ)

scyphistoma (sī-fis′-tȯ-mạ)

Scyphozoa (sī-fȯ-zō′-ạ)

Scytalopus (sī-tal′-ȯ-pus)

Scytonema* (sī-tȯ-nē′-mạ)

sebaceous (sȇ-bā′-shus)

sebific (sȇ-bif′-ik)

sebum (sē′-bum)

Secale* (sēk-ā′-lē, sek-ā′-lē)

secalinus (sek-a-lī′-nus) resembling rye.

Secalis* (sē′-kal-is)

Sechium* (sē′-ki-um)

secretory (sē-krē′-tȯr-i)

sectatrix (sek-ta′-triks) a female follower.

secund (sek′-und)

secundus (sek-un′-dus) second.

sedimentarius (sed-i-men′-tā-ri-us) settling, a sediment.

sedoides (sē-do-ī′-dēz) sedum-like.

Sedum* (sed′-um, sē′-dum)

segetis (sej′-e-tis) of a cornfield.

Seiurus (sī-ū′-rus)

sejugous (sej′-ů-gus) a team of six.

Selaginella* (sel-å-ji-nel′-ạ, sel-ā-jin-el′-ạ)

selaginoides (sel-å-ji-no-ī′-dēz) like *Selago*.

Selago* (sel-ā′-gō)

Selasphorus (sel-as′-fô-rus)

Selenarctos (sel-ēn-ârk′-tos)

Selenodon (se-lē′-nô-don)

Seleucides (sel-ū′-si-dēz)

Selinocarpus* (sel-i-nô-kâr′-pus)

Selinum* (sel-ī′-num)

Semecarpus* (sē-mē-kâr′-pus)

Semele (sem′-e-lē)

semen (sē′-men, pl. sem′-i-nạ)

semidecandrus (sem-i-de-kan′-drus) with half of
    ten stamens.

seminal (sem′-i-nal)

seminiferous (sem-i-nif′-er-us)

Semotilus (sē-mot′-i-lus)

sempervirens (sem-pêr′-vi-renz) evergreen.

Sempervivum* (sem-pêr′-vi-vum, sem-per-vī′-
    vum)

Senebiera* (sen-eb-ē′-rạ)

Senecio* (sē-nē′-shi-ō, sē-nē̂′-si-ō)

senile (sē′-nil, sē′-nīl)

senticosus (sen-ti-kō′-sus) full of thorns.

sentus (sen′-tus) thorny, rough.

sepal (sē′-pal, sep′-al)

Sepedon (sep′-e-don)

Sepiola (sē-pi'-ŏ-lạ)

sepium (sē'-pi-um) of hedges or fences.

Sepside (sep'-si-dē)

septangularius (sep-tan-gū-lā'-ris) corner of a hedge, a fence angle.

septentrionalis (sep-ten-tri-ŏ-nā'-lis) northern, of the north.

septic (sep'-tik)

septum (sep'-tum)

Seraphyta* (sē-raf'-it-ạ)

sere (sēr)

Serenoa* (ser-ên-ō'-ạ)

Sergestes (sêr-jes'-tez)

Sergiolus (ser-ji'-o-lus)

sericatus (sē-rik-ā'-tus) clothed in silks.

Sericidae (sē-ris'-i-dē)

Sericocarpus* (ser-i-kŏ-kâr'-pus, sē-rik-ok-âr'-pus)

Sericostomatidae (ser-i-kŏ-stō-mat'-i-dē, sē-rik-ŏ-stō-mat'-i-dē)

Sericulus (sê-rik'-û-lus)

seriema (ser-i-ē'-mạ)

series (sēr'-ēz, sē'-ri-ēz, pl. sēr'-ēz or sē'-ri-ēz)

Serinus (sē-rī'-nus)

Seriphium* (ser-ī'-fi-um)

serotinus (sē-rō'-ti-nus) late, late-ripe, backward.

Serphidae (sêr'-fi-dē)

Serpula (sêr'-pu-lạ)

serpyllifolius (sêr-pi-li-fol'-i-us, sêr-pi-li-fō'-li-us) thyme-leafed.

Serpyllum* (sêr-pil'-um)

serrate (ser'-āt)

Serratula* (ser-rat'-ŭ-lạ, ser-ā'-tul-ạ)

Serinus. Generic name of the canary. New L. *serinus* <Fr. *serin*, a kind of
bird, a canary. Pronounced: sē-rī'-nus, not ser'-in-us.

Sertularia (sêr-tŭ-lā'-ri-ạ)
serum (sē'-rum)
Sesamum* (ses'-ȧ-mum, sē'-sam-um)
Sesiidae (sēs-ī'-id-ē)
Seseli* (ses'-e-lī)
sessile (ses'-il)
Sesuvium* (sê-sū'-vi-um, ses-ū'-vi-um)
seta (sē'-tạ, pl. sē'-tē)
setaceus (sê-tā'-se-us) bristly, with bristles.
Setaria* (sē-tā'-ri-ạ)
Setochalcis (sêt-ô-kal'-sis)
Setophaga (sē-tof'-ȧ-gạ)
setula (set'-ŭ-lạ) a small bristle.
sexangularis (seks-an-gŭ-lā'-ris) six-angled.
Shibataea* (shib-at'-ê-ạ)
Sialia (sī-ā'-li-ạ)
Sialidae (sī-al'-i-dē)
Sialis (sī'-ȧl-is)
Sibbaldus (sib-al'-dus)
Sibiraea* (sib-ī-rē'-ạ)

Sibynophinae (sib-in-ōf'-in-ē)
siculus (sik'-u-lus) of Sicily.
Sicydium* (sis-id'-i-um)
Sicyos* (sis'-i-os)
Sicyosperma* (sis-i-os-pêr'-mạ)
Sida* (sī'-dạ)
Sidalcea* (sī-dal'-se-ạ)
Sideritis* (sid-ē-rī'-tis)
sierozem (syer'-ŏ-zem)
Sigalphus (sig-al'-fus)
Sigmodon (sig'-mŏ-don)
Signiphoridae (sig-ni-phôr'-i-dē)
Sika (sē'-kạ)
Silaus* (sī-lā'-us)
Silene* (sī-lē'-nē)
siliqua (sil-ik'-wạ)
Siliquaria (sil-i-kwā'-ri-ạ)
siliquastrum (sil-i-kwas'-trum)
Silphidae (sil'-fi-dē)
Silphium* (sil'-fi-um)
Silvanus (sil-vān'-us)
silvaticus (sil-vā'-ti-kus) forest loving.
silvestris (sil-ves'-tris) belonging to a wood or forest.
Silybum* (sil'-i-bum)
Simaruba* (sim-ar-ū'-bạ)
Simenchelyidae (sim-eng-ke'-lī-i-dē)
Simethis* (sim-ē'-this)
simian (sim'-i-an)
simiolus (sī-mi'-ol-us) a little ape.
simplex (sim'-plekz) simple, unmixed, plain.
Simuliidae (sim-ŭ-lī'-i-dē)

simultaneous (sī-mul-tā′-ne-us, sim-ul-tā′-nê-us)
Sinanthropus (sin-an-thrô′-pus)
Sinapis* (sin-ā′-pis)
Sinapus* (sin-ā′-pus)
sinew (sin′-ū)
Sinningia* (sin-inj′-i-ą)
sinuate (sin′-ū-āt)
sinuatus (sin-u-ā′-tus) bent, curved.
sinuous (sin′-û-us)
sinus (sī′-nus, pl. sī′-nus or sī′-nus-ez)
Siphateles (sif-at′-e-lēz)
Siphlonuridae (sif-lon-ûr′-i-dē)
Siphneus (sif′-nê-us)
Sipho* (sif′-ō)
Siphonaptera (sī-fô-nap′-têr-ą)
Siphoniopsis* (sī-fô-ni-op′-sis)
Siphonocladales* (sī-fô-nô-kla-dā′-lēz)
siphonoglyph (sī-fō′-nô-glif)
siphuncle (sī′-fung-k′l)
Siren (sī′-rēn)
Sirenida (sī-ren′-i-dē)
Sirex (sī′-reks)
Siricidae (si-ris′-i-dē)
Sirium* (sī′-ri-um)
Sison* (sī′-son)
Sistrurus (sis-trū′-rus)
Sisymbrium* (sis-im′-bri-um)
Sisyra (sis-ī′-rą)
Sisyridae (sis-ir′-i-dē)
Sisyrinchium* (sis-ir-in′-ki-um)
Sitaris (sit′-à-ris)
sitiens (sit′-i-enz) drying up, thirsty.

Sitodrepa (sit-od-rē'-pạ)
Sitotroga (sit-ot-rō'-gạ)
sitotropism (sī-tot'-rō-pizm)
Sium* (sī'-um)
Skimmia* (skim'-i-ạ)
skolex (skō'-leks)
skotoplankton (skot-ō-plangk'-ton)
sloth (slōth, sloth)
Smeeana* (smē-ā'-nạ)
smegma (smeg'-mạ)
Smicra (smī'-krạ)
Smicrips (smī'-krips)
Smilacina* (smī-lạ-sī'-nạ, smī-las'-in-ạ)
Smilax* (smī-laks)
Smilodon (smī'-lȯ-don)
Smynthuridae (smin-thūr'-i-dē)
Smyrnium* (smûr'-ni-um)
soboles (sob'-ȯ-lēz)
soboliferous (sob-ȯ-lif'-er-us)
Sobralia* (sob-ral'-i-ạ)
sobrinus (sō-brī'-nus) a cousin.
sobrius (sō'-bri-us) not drunk; also, reasonable.
socies (sō'-shi-ēz)
sodalis (so-dā'-lis) a mate, a companion.
Solandra* (sō-lan'-drạ)
Solanum* (sō-lā'-num)
Soldanella* (sol-dan-el'-ạ)
Solea* (sō'-le-ạ)
Solenanthus* (sō-lê-nan'-thus)
Solenobia (sō-lê-nob'-i-ạ)
solenocytes (sō-lē'-nȯ-sīts)
Solenomya (sō-lê-nȯ-mī'-ạ)

Solanum <L. *solanum*, the night-shade. The *o* is long as also is *a*. Pronounced: sō-lā'-num, not sō-lan'-um.

Solidago* (sol-id-ā'-gō)

solidus (sol'-i-dus) firm, dense, not hollow.

Sollya* (sol'-i-ạ)

solstitialis (sol-sti-shi-ā'-lis) belonging to the summer solstice, of or belonging to midsummer.

solutus (sol-ū'-tus) free, loose.

soma (sō'-mạ)

Somateria (sō-mat-ēr'-i-ạ)

somatic (sō-mat'-ik)

somatogenesis (sō-mat-ȯ-je'-ne-sis)

somatopleure (sō'-mat-ȯ-plūr)

somnifer (som'-ni-fêr) bringing sleep.

Sonchus* (song'-kus)

Sophia* (sof'-i-ạ)

Sophora* (sof-ō'-rạ)

Sophronitis* (sof-ron-ī'-tis)

soporator (sop-ȯ-rā'-tor) a sleeper.

soporific (sō-pȯ-rif'-ik, sop-ȯ-rif'-ik)

Sopubia* (sop-ū'-bi-ạ)

Sorbus* (sôr'-bus)

soredium (sō-rē'-di-um)

Sorex (sō'-reks)
Sorocephalus* (sō-rô-sef'-al-us)
sorus (sō'-rus)
Sotalia (sō-tā'-li-ạ)
Sotol (sō'-tōl)
spadices (spā-di-sēz, pl. of spā-diks)
spadonius (spad-ō'-ni-us) barren, seedless.
Spadostyles* (spā-dos'-til-ēz)
Spalax (spā'-laks)
Sparaison (spâr-ȧ-īs'-on)
Sparganium* (spâr-gan'-i-um, spâr-gā'-ni-um)
Spartina* (spâr'-ti-nạ, spar-tī'-nạ)
sparverius (spâr-ver'-i-us) pertaining to a sparrow.
Spatangus (spā-tan'-gus)
spathe (spāth, th as in *those*)
Spathodea* (spath-o'-dê-ạ)
Spathoglottis* (spath-og-lot'-is, spath-og-lō'-tis)
spathose (spā'-thōs)
spathula (spath'-u-lạ) a spatula.
Spathyema* (spath-ī-ē'-mạ)
species (spē'-shēz, spē'-shi-ēz, pl. spē'-shēz)
speciosus (spe-si-ō'-sus) good looking.
Specularia* (spek-û-lā'-ri-ạ)
spelaeology (spē-lê-ol'-ô-ji)
Spelerpes (spē-lêr'-pēz)
speltus (spel'-tus) a kind of wheat.
Speotyto (spê-ot'-i-tō)
Spergula* (spêr'-gŭ-lạ)
Spergularia* (spêr-gŭ-lā'-ri-ạ)
spermaceti (spêr-mȧ-sē'-ti)
Spermacoce* (spêr-mȧ-kō'-sē)
spermatic (spêr-mat'-ik)

spermatid (spêr'-mȧ-tid)
spermatium (spêr-mā'-shi-um)
spermatogonium (spêr-mat-ȯ-gō'-ni-um)
spermatocyst (spêr'-mat-ȯ-sist)
spermatocyte (spêr'-mat-ȯ-sīt)
Spermatophyta (spêr-ma-tof'-i-tạ)
spermatophyte (spêr'-mat-ȯ-fīt)
spermatozoa (spêr-mat-ȯ-zō'-ạ)
spermiducal (spêr-mi-dū'-kal)
Spermolepis* (spêr-mol'-ep-is)
Speyeria (spā-ē'-ri-ạ)
sphacelate (sfas'-e-lāt)
Sphacele* (sfas'-el-ē)
sphactes (sfak'-tēz) a slayer.
Sphaeralcea* (sfē-ral'-sē-ạ)
sphaerocephalus (sfē-rȯ-se'-fa-lus) round-headed.
Sphaerites (sfē-rī'-tēz)
Sphaerocera (sfē-ros'-er-ạ)
Sphecidae (sfes'-i-dē)
Sphecius (sfe'-si-us)
Spheniscus (sfē-nis'-kus)
Sphenodesma* (sfē-nod-es'-mạ)
Sphenodon (sfē'-nȯ-don)
Sphenogyne* (sfē-noj'-in-ē)
Sphenopholis* (sfê-nof-ȯl'-is)
Sphenophorus (sfê-nof'-ȯr-us)
Sphyrapicus (sfī-ra-pī'-kus)
Spica* (spī'-kạ)
spicatus (spī-kā'-tus) having spikes, putting forth
   or having ears or points.
Spicillaria* (spī-sil-ā'-ri-ạ)
Spilanthes* (spī-lan'-thēz)

Spilogale (spī-log′-à-lē)
Spilornis (spī-lor′-nis)
Spilotes (spī-lō′-tēz)
Spilonota (spī-lon-ōt′-ạ)
Spinacia* (spin-ā′-shi-ạ, spī-nā′-si-ạ)
Spindus (spin′-dus)
Spinifex* (spī′-nif-eks)
spinosior (spīn-ō′-si-ôr) more spiny.
spinosissimus (spī-nŏ-sis′-i-mus) very spiny, most spiny.
spinosus (spī-nō′-sus) full of spines or thorns.
spinule (spin′-ūl)
spinulose (spin′-ŭ-lōs, spīn′-ŭ-lōs)
Spinus (spī′-nus)
spiracle (spī′-rà-kl, spir′-à-kl)
Spiraea* (spī-rē′-ạ)
Spiranthes* (spī-ran′-thēz)
Spirodela* (spī-rŏ-dē′-lạ)
Spirontocaris (spī-ron-tok′-âr-is)
Spirostemon* (spī-ros-tē′-mon)
Spirotrichonympha (spī-rŏ-trik-ŏ-nim′-fạ)
Spirula (spir′-ŭ-lạ)
spithameous (spith-ā′-mê-us)
Spiza (spī′-zạ)
Spizella (spī-zel′-ạ)
splenetic (splê-net′-ik, splen′-e-tik)
Spondias* (spon′-di-as)
Spondylus (spon′-di-lus)
sponsalis (spon-sā′-lis) of or belonging to betrothal.
Sporobolus* (spor-ob′-ol-us)
sporogony (spor-oj′-ŏ-ni)

Sporophila (spor-of'-il-ạ)

Spraguea* (sprā'-ge-ạ)

spretus (sprē'-tus) despised, held in contempt.

spumescent (spû-mes'-ent)

spurius (spûr'-ri-us) false, of illegitimate birth.

Spyridium* (spir-id'-i-um)

squalidus (skwā'-li-dus) stiff, rough, foul, unculti-
vated.

squamose (skwā'-mōs)

squarrosus (skwā-rō'-sus) rough, scabby.

squirrel (skwûr'-el)

Stachyopogon* (stak-i-op-ō'-gōn)

Stachys* (stak'-is, stā'-kis)

Stachytarpheta* (stak-it-ar-fē'-tạ)

stagninus (stag-nī'-nus) growing in swampy places.

stalk (stok)

stamen (stā'-men, pl. stā'-menz)

Stangeana* (stanj-e-ā'-nạ)

stans (stanz) erect, upright.

Stapelia* (stā-pel'-i-ạ, stā-pē'-li-ạ)

Staphylea* (staf-i-lē'-ạ)

Staphylinidae (staf-i-lin'-i-dē)

stasis (stā'-sis)

Statice* (stat'-i-sē)

statocyst (stat'-ô-sist)

Staurotypus (stô-rot'-i-pus)

Steatornis (stē-ȧ-tôr'-nis)

steganopodes (steg-ȧ-nop'-ô-dēz)

Steganopus (steg-an'-ô-pus)

Stegnogramme* (steg-nog-ram'-ē)

Stegocephalia (steg-ô-se-fā'-li-ạ)

Stegodon (steg'-ô-don)

Steironema* (stī-rȯ-nē'-mạ)
stele (stē'-lē, pl. stē'-lēz)
Stelephuros* (stel-ef-ū'-ros)
Stelgidopteryx (stel-jid-op'-ter-iks)
Stelis (stē'-lis)
Stellaria* (stel-ā'-ri-ạ)
stellaris (stel-ā'-ris) starry.
stellatus (stel-ā'-tus) starred, covered with stars.
stelligerus (stel-i'-je-rus) bearing stars.
Stellula (stel'-ū-lạ)
stellulatus (stel-ū-lā'-tus) star-like.
Stemodia* (stē-mō'-di-ạ)
Stemonitis* (stem-ȯn-ī'-tis)
Stenanthium* (sten-an'-thi-um)
Stenia* (sten'-i-ạ)
Steno (sten'-ō)
Stenochilus* (sten-ok-ī'-lus)
Stenoglottis* (sten-og'-lot-is, sten-og-lō'-tis)
Stenolobium* (sten-ȯ-lō'-bi-um, sten-ȯ-lob'-i-um)
Stenomesson* (sten-ȯm-es'-on)
Stenomylus (sten-ȯ-mī'-lus)
Stenopelmatidae (sten-ȯ-pel-mat'-i-dē)
Stenorrhina* (sten-ȯ-rīn'-ạ)
Stenosiphon* (sten-os-ī'-fon)
Stenosolenium* (sten-o-sō-lē'-ni-um)
Stenotaphrum* (sten-ot-af'-rum)
Stenotrema (sten-ȯ-trē'-mạ)
Stenotus* (sten'-ȯ-tus)
Stenus (sten'-us, stēn'-us)
Stephania* (stef-ā'-ni-ạ)
Stephanidae (stef-an'-i-dĕ)
Stephanomeria* (stef-an-om-er'-i-ạ)

Stephanophysum* (stef-an-of-ī′-sum)
steppe (step)
Stercorarius (ster-kô-rā′-ri-us)
Sterculia* (ster-kul′-i-ą, ster-kū′-li-ą)
Stereochilus (ster-e-ok-īl′-us)
sterigma (stē-rig′-mą, pl. stē-rig′-ma-tą)
Sterigmostemon* (stē-rig-mos-tē′-mon)
sterilis (ster′-i-lis) sterile, bare, empty.
Steriphoma* (ster-if-ō′-mą)
Sternotherus (ster-nô-thē′-rus)
Stevia* (stē′-vi-ą)
Stibasia* (stib-ā′-shi-ą, stib-ā′-si-ą)
Stichotrematidae (sti-kô-trē-mat′-i-dē)
stigma (stig′-mą, pl. stig′-mat-ą)
stigmata (stig′-mat-ą)
Stilbeae* (stil′-bê-ē)
stilbius (stil′-bi-us) shining.
Stilbopterygidae (stil-bô-te-ri′-ji-dē)
Stipa* (stī′-pą)
stipel (stī′-pel)
stipellate (stī-pel′-āt)
stipes (stī′-pēz), pl. stip′-i-tēz)
stipitate (stip′-i-tāt)
stipularis (stip-ul-ā′-ris) having stipules, pertaining
    to stipules.
stipule (stip′-ūl)
Stipulicida* (stip-ûl-is′-id-ą)
Stizolobium* (stiz-ol-ob′-bi-um, stī-zô-lob′-i-um)
Stizostedion (stī-zô-stē′-di-on)
Stizus (stī′-zus)
Stobartiana* (stob-ârt-i-ā′-ną)
stolidus (stol′-i-dus) unmovable, dull, stupid.

stoma (stō'-mạ, pl. stō'-mat-ạ)
stomadeum (stȯ-mạ-dē'-um)
stomata (stō'-mat-ạ)
stomatic (stōm'-at-ik)
Stomatolepas (stȯ-ma-tȯ-lē'-pas)
stomodaeum (stō-mō-dē'-um, stom-ȯ-dē'-um)
Stomoxyidae (stō-moks'-i-dē)
strabismus (strab-iz'-mus)
stramineous (stram-in'-e-us)
Stratiomyidae (strat-i-ȯ-mī'-i-dē)
Stratiotes* (strat-i-ō'-tēz)
stratum (strā'-tum, pl. strā'-tạ)
Strebla (streb'-lạ)
Strelitzia* (strel-itz'-i-ạ)
Strepsiceros (strep-sis'-er-os)
Strepsiptera (strep-sip'-têr-ạ)
Streptanthus* (strep-tan'-thus)
Streptopelia (strep-tȯ-pē'-li-ạ)
Streptopus* (strep'-top-us, strep'-tȯ-pus)
Streptosolen* (strep-tos-ō'-len)
striatus (stri-ā'-tus) striped, having faint parallel
ridges or lines.
strictus (strik'-tus) drawn tight, pressed together.
striga (strī'-gạ, pl. strī'-jē)
strigatus (strig-ā'-tus) furrowed.
Striges (strī'-jēz)
Strigidae (stij'-i-dē)
Strigilia* (strij-il'-i-ạ)
strigilis (strij'-il-is)
strigosus (strig-ō'-sus) lean, thin, meager.
striola (strī-ōl'-ạ)
Strix (striks)

strobila (strob'-i-lạ, strȯ-bī'-lạ)
Strobilanthes* (strob-il-an'-thēz)
strobile (strob'-il, strōb'-īl)
Strobilorachis* (strob-il-ȯr'-ak-is)
strobilus (strob'-i-lus, pl. strob'-i-lē)
Strobus* (strob'-us, strō'-bus)
Stromatopora (strō-ma-top'-ȯ-rạ)
Strophostyles* (strof-ȯ-stī'-lēz)
struma (strū'-mạ)
strumatus (strū-mā'-tus) with tumors.
strumosus (strū-mō'-sus) scrofulous, swellen.
Struthio (strū'-thi-ō)
Struthiomimus (strū-thi-ȯ-mim'-us)
Struthium (strū'-thi-um)
strychnine (strik'-nin)
Strychnos* (strik'-nos)
Strymon (strī'-mon)
stupulosus (stȗ-pū-lō'-sus) covered with fine, short
    hairs.
Stylandra* (stī-lan'-drạ)
Stylocline* (stī-lok-lī'-nē)
Stylolepis* (stī-lol'-ep-is)
Stylonychia (stī-lȯ-nik'-i-ạ)
Stylophorum* (stī-lof'-ȯr-um)
Stylopidae (stī-lop'-id-ē)
Stylosanthes* (stī-lȯs-an'-thēz)
Styphelia* (stif-el'-i-ạ)
styraciflua (stir-ak-if'-lu-ạ)
Styracosaurus (stir-rak-ȯ-sȏ'-rus, stī-rak-ȯ-sȏ'-rus)
Styrax* (stir'-aks, stī'-raks)
Suaeda* (sȗ-ē'-dạ, swē'-dạ)

suaveolens (swā-ve'-o-lenz) sweet-scented, fragrant.

subaerial (sub-ā̇-ē'-ri-al)

subanconeus (sub-ang-kōn'-e-us)

Suber* (sū'-bêr)

suberectus (sub-ē-rek'-tus) raised up.

suberin (sū'-bêr-in)

Suberites (sū-bêr-ī'-tēz)

suberosus (sūb-ē-rō'-sus) corky in texture.

subitus (sub'-i-tus) sudden, unexpected.

submersus (sub-mêr'-sus) submerged, plunged under water.

subruficollis (sub-rū-fi'-kol-is) red under the neck.

subsequus (sub-se'-kwu-us)

subsidence (sub-sīd'-ens, sub'-si-dens)

substratum (sub-strā'-tum)

subterraneus (sub-ter-ā'-ne-us) beneath the ground.

Subularia* (sū-bû-lā'-ri-ạ)

subulate (sū'-bû-lāt)

subulatus (sū-bû-lā'-tus) awl-like, thread-like and tapering to a fine point.

subulicorn (sū'-bū-li-kôrn)

Succisa* (suk-sī'-sạ)

succisus (suk-sīs'-us) cut off, imasculated, made powerless.

Sueda* (swā'-dạ)

suffocatus (suf-ō-kā'-tus) suffocated, strangled.

suffruticose (suf-rū'-ti-kōs)

suinus (su-īn'-us) belonging to swine.

sula (sū'-lạ)

sulphurea (sul-fûr'-e-ạ)

sumac (sū'-mak, shōō'-mak)
Sunipia* (sū-nī'-pi-ạ)
supinator (sū-pi-nā'-tôr)
supine (sū'-pīn)
supinus (sup-īn'-us) lying on the back, bent back.
surculaceus (sûr-ku-lā'-se-us) woody, like wood.
surculatus (sûr-kū-lā'-tus) trimmed, pruned.
surculose (sûr'-kŭ-lōs)
surculosus (sûr-kŭ-lō'-sus) woody, like wood.
Suriana* (sū-ri-ā'-nạ)
Suricata (sū-ri-kā'-tạ)
surinam (sū'-ri-nam)
Surnia (sûr'-ni-ạ)
Sus (sus)
Susarium* (sū-sar'-i-um)
Sutrina* (sū-trī'-nạ)
suturalis (sūt-ū-rā'-lis) pertaining to a seam.
Swertia* (swêr'-ti-ạ)
sycon (sī'kon)
syconium (sī-kō'-ni-um)
Syctodes (sik-tō'-dēz)
sylvaticus (sil-vā'-ti-kus) growing in woods.
Sylvicapra (sil-vi-kap'-rạ)
Sylviidae (sil-vī'-i-dē)
symbiosis (sim-bī-ō'-sis)
symbiotic (sim-bī-ot'-ik)
Symphachne* (sim-fak'-nē)
Sympherobiidae (sim-fe-rŏ-bī'-i-dē)
Symphoricarpus* (sim-fō-ri-kâr'-pus)
Symphyandra* (sim-fi-an'-drạ)
Symphyla (sim'-fi-lạ)
symphysis (sim'-fi-sis)

Symphytum* (sim′-fit-um)
Sympieza* (sim-pi-ez′-ạ, sim-pi-ēz′-ạ)
Symplocarpus* (sim-plok-âr′-pus)
Symplocos* (sim′-plok-os)
Synandra* (sin-an′-drạ)
synapse (sin-aps′)
synapsis (sin-ap′-sis)
Synaptomys (sin-ap′-tȯ-mis)
Synarrhena* (sin-âr′-ren-ạ)
Syncarida (sin-kâr′-id-ạ)
Synceros (sin′-se-ros)
synconium (sin-kō′-ni-um)
syncytium, (sin-sish′-i-um, sin-sit′-i-um)
syndesis (sin-dē′-sis)
synergid (sin-êr′-jid)
Synetheres (sin-eth′-ė-rēz)
Syngenesia* (sin-jen-es′-i-ạ)
Syngonium* (sing-ō′-ni-um)
synhesma (sin-hes′-mạ)
Synlestidae (sin-les′-ti-dē)
synoekete (sin-ē-kēt′)
Synsiphon* (sin′-sif-ōn)
Syntelia (sin-tē′-li-ạ)
syntenosis (sin-te-nō′-sis)
Synthliboramphus (sin-thli-bȯ-ram′-fus)
Synthyris* (sin-thī′-ris, sin′-thi-rus)
Syntoechus (sin-tē′-kus)
Syntrichopappus* (sin-trik-ȯ-pap′-us)
Syringa* (sī-rin′-gạ, si-ring′-ạ)
syrinx (sir′-ingks, pl. sir′-in-jēz)
Syrphidae (sûr′-fi-dē)
Syrrhaptes (sir-rap′-tēz)

syssarcosis (sis-âr-kō'-sis)
systemic (sis-tē'-mik, sis-tem'-ik)
Systena (sis'-tê-nạ)
systole (sis'-tō-lē)
sistylus (sis-tī'-lus) with styles together.
Syzygium* (siz-ij'-i-um)

T

tabaccarius (tab-ak-ā'-ri-us) pertaining to tobacco;
    also, a pipe for smoking.
tabacinus (tab-as'-i-nus)
Tabanidae (tab-an'-i-dē)
Tabanus (tab-ā'-nus)
Tabebuia* (tab-eb-ū'-i-ạ, ta-be-bû-ĭ'-ạ)
tabescens (tā-bes'-enz) decaying, wasting, away.
tabidus (tā'-bid-us) decaying, corrupting.
Tachibaptes (tak-i-bap'-tēz)
Tachina (tǎ'-kin-ạ)
Tachinidae (tǎ-kin'-i-dē)
Tachycineta (tak-i-sin-ēt'-ạ)
Tachyporus (tak-ip'-ôr-us)
Tadarida (ta-da'-ri-dạ)
Taenia (tē'-ni-ạ)
taeniatus (tē-ni-ā'-tus) banded.
taeniiform (tē-ni'-i-fôrm)
Tagelus (tā'-je-lus)
Tagetes (tā-jē'-tēz)
taiga (tī'-gạ)
Talegallus (tal-ê-gal'-us)
Talinopsis* (tal-in-ops'-is)
Talinum* (tal-i'-num)
Talpa (tal'-pạ)

Tamandua (tam-an'-dŭ-ą)
Tamarindus* (tam-ar-in'-dus)
Tamarix* (tam'-âr-iks)
Tamias (tā'-mi-as)
Tamus* (tā'-mus)
tanacetifolius (tan-ȧ-sē-ti-fol'-i-us, tan-ȧ-sē-ti-fō'-li-us) tansy-leafed.
Tanacetum* (tan-ȧ-sē'-tum)
Tanaecium (ta-nē'-si-um)
Tanais (tā-nā'-is)
Tangavius (tan-gā'-vi-us)
Tantilla (tan-tē'-yą)
Tanypus (tan'-ip-us)
Tapacolas (tap-ȧ-kōl'-as)
tapetum (tap-ēt'-um)
Taphonycteris (taf-ô-nik'-ter-is)
Taphozous (taf'-ô-zō'-us)
Taphrina* (taf-rī'-ną)
Tapinoma (tap-i-nō'-mą)
tapir (tāp'-êr)
Tapirus (tap'-ir-us)
Tapogomea* (tā-pog-ō'-me-ą)
tarandrus (tar-an'-drus) an animal of northern countries.
Taraxacum* (tar-ak'-sȧ-kum)
Tardigrada (târ-dig'-rad-ą)
Tarenna* (târ-en'-ą)
Tarsipes (târ'-si-pēz)
Tarsius (târ'-si-us)
Tarsonemus (târ-sô-nē'-mus)
Tasmania* (tas-mān'-i-ą)
Tatarida (ta-târ'-id-ą)

Tatusia (ta-tū'-si-ạ)
Taurotragus (tô-rọ̄-trāg'-us)
Taxidea (tak-sid'-ê-ạ)
Taxodium* (tak-sō'-di-um)
Taxus* (tak'-sus)
Tchitrea (chi'-tre-ạ)
Tecoma* (tê-kō'-mạ, tek-ō'-mạ)
tectorum (tek-tō'-rum) of a roof, of a cover.
tegenaria (tej-e-nā'-ri-ạ)
tegens (te'-jenz) covering.
tegula (teg'-ụ̄-lạ)
tegumen (teg'-ụ̄-men)
Teiidae (tē'-i-dē)
Teius (tē'-us)
Telamona (tel-am-ōn'-ạ)
Telea (tē'-lê-ạ)
Telenomus (tê-len'-ọ̄-mus, tel-en'-om-us)
teleology (tel-ê-ol'-ọ̄-ji, tē-lē-ol'-ọ̄-ji)
Telephorus (tel-ef'-ôr-us)
Telipogon* (tē-lip-ō'-gōn)
telium (tē'-li-um, tel'-i-um)
Tellina (te-lī'-nạ)
Telmatodytes (tel-mat-ọ̄-dī'-tēz)
telolecithal (tel-ọ̄-les'-ith-al)
Telopea* (tē-lō'-pe-ạ)
telophase (tel'-ọ̄-fāz)
telotarsus (tel-ọ̄-târ'-sus)
Telphusa (tel-fū'-sạ)
telum (tē'-lum) a weapon, a missle.
temulentus (tē-mu-len'-tus) drunk, inebriated.
Temenuchus (tem-e-nū'-kus)
temperature (tem'-pêr-ạ-tûr)

tenaculum (ten-ak'-u-lum)

Tenaga (ten-ā'-gạ)

Tenaris* (tē'-nar-is)

tenax (ten'-aks) holding fast, tight, firm.

Tendana* (ten-dā'-nạ)

Tenebrionidae (tĕ-neb-ri-on'-i-dē)

tenebrosus (ten-ĕ-brō'-sus) dark, gloomy.

tenellus (ten-el'-us) somewhat tender or delicate.

teneral (ten'-êr-al)

Tenodera (ten-od'-er-ạ)

Tenthredo* (ten-thrē'-dō)

tentorium (ten-tō'-ri-um)

tenuiflorus (ten-û-i-flō'-rus) thin-flowered, weak-flowered, slender-flowered.

tenuifolius (ten-û-i-fol'-i-us, ten-û-i-fō'-li-us) thin- or weak-leafed, slender-leaved.

tenuipes (ten-ū'-i-pēz) weak-footed.

tenuis (ten'-û-is) thin, lank; also, weak.

tenuissimus (ten-û-is'-i-mus) most weak or thin.

tepal (tep'al)

Tephroclamys (tef-rok'-lam-is)

Tephritis (tef-rīt'-is)

Tephroclystis (tef-rŏ-klis'-tis)

tephrosanthus (tef-ros-an'-thus) with ash-colored flowers.

Tephrosia* (tef-rō'-shi-ạ, tef-rō'-si-ạ)

Teracolus (ter-ak'-ol-us)

Terapene (ter-ạ-pē'-nē)

Teras (tē'-ras)

Terathopius (ter-à-tho'-pi-us)

Terebra (ter'-eb-rạ)

Teredo (ter-ē'-dō)

Terekia (ter-ek′-i-ạ)

teres (tē′-rēz)

terete (tē-rēt′, ter′ēt)

Teretistris (ter-ēt-is′-tris)

teretiusculus (ter-ēt-i-us′-ku-lus) almost smooth, well-rounded, cylindrical.

tergesus (ter′-ges-us) polished.

tergite (ter′-jīt, ter′-gīt)

tergum (ter′-gum, têr′-gum)

Terminalia* (têr-min-ā′-li-ạ)

Termitidae (têr-mit′-i-dē)

Termitoxeniidae (têr-mit-ȯ-zen-ī′-i-dē)

Ternatea* (ter-nā′-te-ạ)

ternatus (ter-nā′-tus) consisting of three.

ternipes (ter′-ni-pēz)

Ternstroemia* (têrn-strē′-mi-ạ)

Terpsiphone (terp-si-fō′-nē)

terrestris (ter-es′-tris) belonging to the earth.

Tertiary (têr′-shi-â-ri)

Tesia (tē′shi-ạ, tē′si-ạ)

tesotus (tes-ō′-tus) stiff, hard, firm.

Tessaria* (tes-ā′-ri-ạ)

tesselatus (tes-el′-ā-tus) of small stone, checkered.

Tetanocera (tet-an-os′-er-ạ)

teter (tē′-ter) offensive, foul, loathsome.

Tethys (tē′-this)

Tetrabelodon (tet-rȧ-bel′-ȯ-don)

Tetracera* (tet-ras′-er-ạ)

Tetracha (tet′-rȧ-kạ)

Tetragonia* (tet-rȧ-gō′-ni-ạ)

Tetragonotheca* (tet-rag-ȯ-nȯth-ē′-kạ)

tetragonum (tet-rȧ-gō′-num) a quadrangle.

Tetralix* (tet′ral-iks)
Tetramera (tet-ram′-e-ra̧)
Tetranychus (tet-ran′-i-kus)
tetrancistus (tet-ran-sis′-tus)
Tetrandrus (tet-ran′-drus)
Tetrao (tet′-rå-ō)
Tetraogallus (tet-rå-o̊-gal′-us)
Tetraoperdix (tet-rå-o̊-pêr′-diks)
Tetrapanax* (tet-rap′-an-aks)
tetraploidy (tet′-ra-ploy′-di)
Tetrapogon* (tet-rap-ō′-gōn)
tetraspermus (tet-ra-spêr′-mus) four-seeded.
Tetrastichidae (tet-ra-stik′-i-dē)
Tetrastichus (tet-ra′-stik-us)
Tetrix (tē′-triks)
Tettigidae (tet-ij′-i-dē)
Tettigoniidae (tet-i-gon-ī′-i-dē)
Teucrium* (tū′-kri-um)
thalamus (thal′-a-mus)
Thalarctos (thal-årk′-tos)
Thalasseus (thal-as′-e-us)
Thalassochelys (thal-a̧-sok′-el-is)
Thalia* (tha̅′-li-a̧)
Thalictrum* (thal-ik′-trum)
thalassoid (thal-as′-oyd)
Thallophyta (thal-of′-ita̧)
Thamnophis* (tham′-no̊-fis)
Thamnosma* (tham-nos′-ma̧, tham-noz′-ma̧)
Thanasimus (than-as′-im-us)
Thanatus (than′-a̧-tus)
thlaspiformis (thla-spi-fôr′-mis) of the form of *Thlaspi*.

tharus (thä'-rus)
Thaspium* (thas'-pi-um)
Thaumatoxenidae (thô-mat-ô-zen'-i-dē)
Thea* (thē'-ạ)
Theca* (thek'-ạ)
theca (thē'-kạ)
Thecla (thek'-lạ)
Thecophora (thē-kof'-ô-rạ)
Thecostele* (thē-kos-tē'-lē)
Thelephora* (thē-lef'-ôr-ạ)
Thelesperma* (thē-les-spêr'-mạ)
Thelocactus* (thēl-ô-kak'-tus)
Thelphusa (thel-fū'-sạ)
Thelygonum* (thê-lig'-on-um)
Thelymitra (thê-lim-ī'-trạ)
Thelyphonus (thê-lif'-ô-nus)
Thelypodium* (thē-li-pod'-i-um)
Thelypogon* (thê-lip-ō'-gōn)
Themistoclesia (them-is-tok-lē'-si-ạ)
thenal (thē'-nal)
thenar (thē'-nâr)
Theobroma* (thē-ô-brō'-mạ)
Theraphosa (ther-à-fō'-sạ)
Theraphosidae (ther-à-fos'-i-dē)
therapod (thē'-rà-pod)
Thereva (ther-ēv'-ạ, ther'-e-vạ)
Therevidae (thê-rev'-i-dē)
Theridium (thê-rid'-i-um)
Therina (thê-rī'-nạ)
theriodonta (thē-ri-ō-don'-tạ)
Thermesia (thêr-mē'-shi-ạ, thêr-mē'-si-ạ)
therophyte (ther'-ô-fīt)

Thesium* (thē'-shi-um, thê'-si-um)
thesocytes (thē'-sô-sīts)
Thespesia* (thes-pē'-shi-ą, thes-pēs'-ią)
Thetomys (thēt'-ô-mis)
Thevetia* (thê-vē'-shi-ą, thê-vē'-ti-ą)
thigmotropism (thig-mot'-rô-pizm)
thinobates (thīn-ô-bā'-tēz)
Thinocoridae (thīn-ô-kôr'-i-dē)
Thinocorus (thīn-ok'-ô-rus)
thinophyte (thīn'-ô-fīt)
Thinopus (thīn'-ô-pus)
Thiobacteria* (thī-ô-bak-tē'-ri-ą)
Thlaspi* (thlas'-pī)
Thoe (thō'-ē)
Thomomys (thō'-mô-mis)
thorax (thō'-raks, pl. thō'-rā-sēz)
Thos (thōs)
Threskiornis (thrēs-ki-ôr'-nis, thres-ki-ôr'-nis)
Thrinax* (thrī'-naks)
Thrincia* (thrin'-shi-ą, thrin'-si-ą)
Throscus (thros'-kus)
Thryallis* (thrī-al'-is)
Thryomanes (thrī-ô-mān'-ēz)
Thryospiza (thrī-ô-spī'-zą)
Thryothorus (thrī-oth'-ô-rus)
Thuja* (thū'-ją)
Thujopsis* (thû-jop'-sis)
Thunbergia* (thun-bêr'-gi-ą)
Thuya* (thū'-yą)
thylacine (thī'-là-sin)
Thylacinus (thī-las'-i-nus)
Thylacynus (thī-las'-i-nus)

Thylogale (thī-log′-al-ē)
Thymallus (thī-mal′-us)
Thymus* (thī′-mus)
Thynnidae (thin′-i-dē)
Thyone (thī′-ȯ-nē)
Thyreocoris (thī-rē-ok′-ȯr-is)
Thyreus (thī′-rê̂-us)
Thyridopteryx (thī-rid-op′-têr-iks)
Thyroptera (thī-rop′-têr-ạ)
thyrsiflorus (thêr-si-flō′-rus) with flowers arranged
   in a thyrsis or contracted panicle.
thyrsus (thêr′sus)
Thysanocarpus* (thī-sa-nȯ-kâr′-pus,   this-an-ȯ-
   kâr′-pus)
Thysanoptera (thī-sa-nop′-têr-ạ, thi-sa-nop′-têr-ạ)
Thysanura (thī-sȧ-nū′-rạ, thi-sȧ-nū′-rạ)
Tiarella* (tī-ȧ-rel′-ạ)
tiburon (ti-bū-rōn′)
Tichodroma (tī-kod′-rȯ-mạ)
Tichosurus (tī-kos′-ûr-us)
tige (tīj)
Tigridia* (tī-grid′-i-ạ)
Tigrisoma (tī-gri-sō′-mạ)
Tilia* (til′-i-ạ)
Tillandsia* (til-and′-si-ạ)
Timalia* (tī-mā′-li-ạ)
Timelia (tī-mē′-li-ạ)
Tinamus (tin′-ȧ-mus)
tinctorius (tink-tō′-ri-us) belonging to dyeing; also,
   blood-thirsty.
Tinea (tin′-ê̂-ạ)
Tineidae (ti-nē′-i-dē)

Tineina* (tin-e-ī′-na̧)

Tingidae (tin′-ji-dē)

Tingis (tin′-jis)

Tintinnus (tin-tin′-us)

tinus (tī′-nus) a plant, prob., a *Viburnum*.

Tiphia (tif′-i-a̧)

Tiphiidae (tif-ī′-i-dē)

Tipularia* (tip-ū-lā′-ri-a̧, tip-ul-ā′-ri-a̧)

Tithonia* (ti-thō′-ni-a̧)

Tithymalus* (tith-im′-al-us)

Titragyne* (tit-raj′-in-ē)

Tobira* (tob-ī′-ra̧)

Tococa* (tok-ō′-ka̧)

Todea* (tō′-de-a̧)

Todirostrum (tō-di-ros′-trum)

tokostome (tok′-os-tōm)

Tolmiea* (tŏl-mē′-a̧)

Tolypeutes (tol-i-pū′-tēz)

tomentosus (tō-men-tō′-sus) full of matted hairs, covered with matted hairs.

tomentum (tō-men′-tum)

Tomeutes (tom-ū′-tēz)

Tomex* (tō′-meks)

Tomicus* (tom′-ik-us)

Tomistoma (tom-is′-tŏ-ma̧)

tomium (tō′-mi-um)

Tomocerus (tōm-os′-er-us)

Tomoxia (tōm-oks′-ia̧)

tonotaxis (ton-ŏ-taks′-is)

topotype (top′-ŏ-tīp)

Tordylium* (tôr-di′-li-um)

torminalis (tôr-mi-nā′-lis) good against colic.

torosus (tôr-ō′-sus) full of muscle, lusty.
tortilis (tôr′-til-is) twisted, twined, winding.
tortoise (tôr′-tus, tôr′-tis)
Tortricidae (tôr-tris′-i-dē)
Totanus (tot′-ȧ-nus)
totipotent (tŏt-ip′-ŏt-ent)
towhee (tou′-hē; tō′-hē)
Toxostoma (toks-os′-tō-mạ)

Toxostoma <Gr. *toxon*, a bow+*stoma*, mouth. Generic name of many of the Thrashers which have bowed beaks. Pronounced: toks-ost′-ōm-ạ.

trabecula (trab-ek′-ū-lạ)
trachea (trak-ē′-ạ, trā′-ke-ạ)
Trachelas (trak-ē′-las)
Trachelipoda (trak-ê-lip′-ō-dạ)
Trachelium* (trȧ-kē′-li-um)
Trachelospermum* (trȧ-kēl-os-pêr′-mum)
Trachinus (trȧ-kī′-nus)
Trachymene* (trak-i-mē′-nē)
trachyodon (trak-i′-ō-don) with rough teeth.
Tradescantia* (tra-des-kan′-shi-ạ, tra-des-kan′-ti-ạ)

Traganum* (tra'-gan-um)
Tragelaphus (tra-jel'-á-fus)
Tragia* (traj'-i-ạ)
Tragopan (trag'-ọ-pan)
Tragopogon (trag-ọ-pō'-gōn)
Tragulina (trag-u-lī'-nạ)
Tragulus (trag'-u-lus)
tragus (trā'-gus)
Trametes* (trâ'-met-ēz)
Trapa* (trā'-pạ, trap'-ạ)
Trema (trē'-mạ)
Tremarctos (trê-mârk'-tos)
Trematoda (trē-ma-tōd'-ạ, trem-á-tōd'-ạ)
Tremex (trē'-meks)
tremulus (trem'-u-lus) trembling, that which causes one to tremble.
Treron (trē'-rōn, trē'-ron)
triandrus (trī-an'-drus) three-anthered.
Triblemma (trib-lem'-ạ)
Tribolium (trib-ol'-i-um)
Triboloceratidae (trib-ọ-lō-se-rat'-i-dē)
Trichomonas (tri-kom'-ọ-nas)
triboluminescence (trib-ọ-loo-min-es'-ens)
Tribonyx (trib'-ọ-niks)
Tribrachium* (trī-brak'-i-um)
Tribulus* (trib'-ul-us)
Tricantha* (trik-an'-thạ)
Triceratops (trī-ser'-á-tops)
Trichachne* (trī-kak'-nē)
trichas (trī'-kas) a thrush.
Trichechus (trik'-e-kus)
trichidium (trik-id'-i-um)
Trichilia* (trik-il'-i-ạ)

trichiniasis (trik-in-ī′-ás-is)
Trichobius (trik-ob′-i-us)
Trichodectidae (trik-ô-dek′-ti-dē)
trichoides (trik-o-ī′-dēz) hair-like.
Tricholaena* (trik-ô-lēn′-ạ)
Trichomanes* (trik-om′-á-nēz)
Trichomonas (trik-om′-ô-nas)
Trichonema* (trik-ô-nē′-mạ)
Trichoplusia (trik-op-lū′-si-ạ)
trichopes (trik′-ô-pēz) hairy-footed.
Trichopoda (trik-op′-ôd-ạ)
Trichopteryx (trik-op′-têr-iks)
Trichoptilum* (trik-op-ti′-li-um)
Trichosanthes* (trik-os-an′-thēz)
Trichostema* (trik′-ô-stē-mạ)
Trichosurus (trik-os′-ûr-us)
trichotomous (trī-kot′-ô-mus)
tricolor (trik′-ul-ôr) three-colored.
tricornis (trik-ôr′-nis) three-horned.
Tricyrtis* (trī-sir′-tis)
tridens (trid′-enz)
tridentatus (trid-en-tā′-tus) three-toothed.
Tridymus (trid′-i-mus)
Trientalis* (tri-en-tā′-lis)
Triepeolus (trī-ep-ē′-ô-lus)
trifarious (trif-ā′ri-us)
triferous (trif′-er-us)
trifid (trif′-id)
trifidus (trif′-i-dus) cut into three parts.
trifoliate (trī-fō′-li-āt)
Trifolium* (trif-ol′-i-um, trī-fō′-li-um)
trifurcus (trif-ûr′-kus) three-forked.

Trifolium <L. *trifolium*, trefoil, a "three leaved grass" <*tri* (Gr. *tris*) three times+*folium*, leaf. Pronounced: trif-ol′-i-um. Often pronounced trī-fō′-li-um.

trigamy (trig′-a-mi)
Triglochin* (trig-lō′-kin)
Triglossum* (trig-lō′-sum, trī-glo′-sum)
triglumis (trī-glūm′-is) with three glumes.
Trigonella (trig-o̊-nel′-ạ)
trigonal (trig′-o̊-nal)
trigone (trī′-gon, trī′-gōn)
Trigonia (trig-ō′-ni-ạ)
Trigonocephalus (trig-o̊-no̊-sef′-al-us)
trigonophyllus (trig-o̊-no̊-fil′-us) three-angled leaf.
trigynus (trij′-i-nus) three-pistiled.
trilineata (tril-i-ne-ā′-tạ)
Trilisa* (tril′-i-sạ)
trima (trī′-mạ)
Trimeresurus (trim-er-e-sū′-rus)
trimerous (trim′-er-us)
trimestris (trim-es′-tris)
Trimorphodon (trī-mōrf′-o̊-don)
trinervis (trī-nêr′-vis) three-nerved.
Trinia* (trī′-ni-ạ)
Trinoton (trī-nō′-ton)

Triodia* (tri-ō′-di-ạ, tri-od′-i-ạ)
Triodytes (tri-ō̊-dī′-tēz)
Trionyx (tri′-ō̊-niks)
Triops (tri′-ops)
Triopteris* (trī-op′-ter-is)
Triosteum* (tri-os′-tê-um)
tripartitus (tri-pâr-tī′-tus) divided into three parts.
Tripetalus* (trip-et′-al-us)
triphyllus (trif-il′-us) three-leaved.
Tripidae (trip′-i-dē)
Triplaris* (trip-lā′-ris)
Triplasis* (trip-lās′-is)
triploid (trip′-loyd)
Triplopus (trip′-lō̊-pus)
triplostichous (trip-los′-tik-us)
tripodalis (trip-od-ā′-lis)
Triprocris (trip′-rok′-ris)
Tripsacum* (trip′-sȧ-kum)
triquetrus   (trī-kwê′-trus, trī-kwet′-rus)   three-
   angled.
Trisetum* (tris-ē′-tum, trī-sē′-tum)
Tristania* (tris-tā′-ni-ạ)
tristis (tris′-tis) dejected, miserable.
tristyly (trī-stī′-li)
trisulcus (tris-ul′-kus) three-pointed, triple.
Triteleia* (trit-el-ī′-ạ)
Triteleiopsis* (trit-el-ī-ops′-is)
Triticum* (trit′-i-kum, trī′-tik-um)
Tritoma* (trit′-ō̊-mạ)
Triton (trī′-ton)
Tritonia* (trī-tō′-ni-ạ)
Triturus (trit-ū′-rus)

triumphans (tri-um'-fanz)
triungulin (trī-ung'-gū-lin)
Triuris* (tri-ū'-ris)
trivialis (triv-i-ā'-lis) common, ordinary, found everywhere.
Trixoscelis (triks-os'-sel-is)
trochanter (trō-kan'-têr)
Trochelminthez (trok-hel-min'-thēz)
Trochilus (trok'-il-us)
trochlear (trok'-lē-âr)
Trochocarpa (trok-ȯ-kâr'-pạ)
trochophore (trok'-ȯ-fôr)
Trochotoma (trok-ot'-ȯ-mạ)
Trochus (trō'-kus)
Troctes (trok'-tēz)
Trogidae (troj'-i-dē)
Troglodytes (trōg-lȯ-dī'-tēz, trōg-lȯd'-i-tēz, trog-lȯ-dī'-tēz)
Trogoderma (trō-gȯ-dêr'-mạ)
trogon (trō'-gon)
Trogosita (trō-gȯ-sī'-tạ)
troilus (trō'-i-lus)
Trollius* (trol'-i-us)
Trombidium (trom-bi'-di-um)
Tropaeolum* (trȯ-pē'-ol-um)
Trophianthus* (trof-i-an'-thus)
trophobiosis (trof-ȯ-bī'-ȯ-sis)
trophozoite (trof-ȯ-zō'-īt)
Tropidia* (trop-id'-i-ạ)
Tropidocarpum* (trop-id-o-kâr'-pum)
Tropidoclonion (trop-id-ȯ-klon'-i-on)
Tropidopria (trop-id-ō'-pri-ạ)

Tropidocarpum <Gr. *tropis*, genit. *tropidos*, the keel of a ship+*karpos*, fruit. The initial *o* is short. Pronounced: trop-id-ô-kâr′-pum, not trô-pid-ô-kâr′-pum.

Tropidodipsas (trop-id-ô-dips′-as)
Tropidonotus (trop-id-ô-nô′-tus)
tropism (trô′-pizm)
tropophyte (trop′-ô-fīt)
tropotaxis (trop-ô-tak′-sis)
Trox (troks)
Troximon* (troks′-i-mon)
Trutta (trut′-a̧)
Trygon (trī′-gon)
tryma (trī′-ma̧)
Trypanosoma (trip-a̧-nô-sō′-ma̧)
trypanosome (trip-an′-ô-sōm)
Trypeta (trī-pēt′-a̧)
Trypetidae (trī-pet′-i-dē)
Trypoxylon (trī-pok′-si-lon)
Tsuga* (tsū′-ga̧)
tuberosus (tū-be-rō′-sus) full of humps.
tubula (tub′-u-la̧) a small trumpet.
Tubularia (tub-ŭ-lā′-ri-a̧)
Tulipa* (tū′-lip-a̧)
Tumboa* (tum′-bô-a̧)

tumescent (tū-mes'-ent)
Tunga (tun'-gą)
Tupaia (tū-pā'-yą)
Tupinambis (tup-i-nam'-bis)
Turacus (tū'-rą-kus)
Turbellaria (tûr-bel-ā'-ri-ą)
Turdoides (tûr-do-i'-dēz)
Turdus (tûr'-dus)
turgescent (tûr-jes'-ent)
turgid (tûr'-jid)
turgor (tûr'-gôr)
Turritis* (tûr-ī'-tis)
Tursiops (tūr'si-ops)
Tussilago* (tus-i-lā'-gō)
tylarus (til'-ȧ-rus)
Tyloglossa* (tī-log-los'-ą, tī-log-lō'-są)
tylosis (tī-lō'-sis)
tylosurus (tī-lȯ-sū'-rus)
tylote (tī'-lōt)
Tympanuchus (tim-pȧ-nū'-kus)
tympanum (tim'-pan-um)
Typha* (tī'fą)
Typhlocyba* tif-lok-ī'-bą)
Typhlops (tif'-lops)
typhlosole (tif'-lȯ-sōl)
Typhlotriton (tif-lȯ-trī'-ton)
Typhonium* (tī-fō'-ni-um)
Tyrannosaurus (tī-ran-ȯ-sô'-rus)
Tyrannus (tī-ran'-us)
Tyroglyphus (tī-rog'-li-fus)
Tyto (tī'-tō)

U

ubericolor (ūb-er-i'-ku-lôr) rich in color.
Uca (ōō'-ką)
Udora* (ud-ō'-rą)
uletic (ū-let'-ik)
Ulex* (ū'-leks)
uliginose (ū-lij'-i-nōs)
uliginosus (ū-lij-i-nō'-sus) wet, full of moisture.
Ulmus* (ul'-mus)
ulnare (ul-nā'-rē)
Uloboridae (ū-lob-ôr'-id-ē)
Uloborus (ū-lob'-or-us)
Ulothrix (ū'-lŏ-thriks)
Ulotrichi (ů-lot'-rik-ī)
ulula (u'-lu-lą) a screech-owl.
Ululodes (ul-ul-ō'-dēz)
Ulva* (ul'-vą)
Uma (ū'-mą)
umbellatus (um-bel-ā'-tus) umbelled, with umbels.
Umbellularia* (um-bel-ul-ā'-ri-ą)
umbilical (um-bi-lī'-kal, um-bi'-li-kal)
umbilicus (um-bi-lī'-kus, um-bil'-i-kus)
umbo (um'-bō, pl. um-bŏ'-nēz)
umbonal (um-bō'-nal, um'-bŏ-nal)
umbrinus (um'-brī-nus) darkened, shady.
umbrosus (um-brō'-sus) shady.
Uncinula* (un-sin'-ůl-ą)
uncus (ung'-kus)
undatus (un-dā'-tus) wavy.
undosus (un-dō'-sus) full of waves.
undulatus (un-dul-ā'-tus) wavy, full of waves.
Unedo* (ū'-ned-ō)

Ungnadia* (un-gnā'-di-ạ)
unguiculate (un-gwik'-ụ-lāt)
unguligrade (ung'-ụ-li-grād)
unicolor (ū-nik'-ul-ôr)
uniflorus (ū-ni-flō'-rus) one or single-flowered.
uniglumis (ū-ni-glūm'-is) with a single glume.
unijugate (ū-ni-jū'-gāt)
Uniola*(ū-nī'-ộ-la)
Unisema* (ū-nis-ē'-ma)
unisexual (ū-nis-eks'-u-al)
univalent (ū-niv'-al-ent, ūn-i-vāl'-ent)
Upupa (ū'-pŭ-pạ, u'-pu-pạ)
urachus (ū'-rak-us)
Uralepsis* (ū-ral-ep'-sis)
Urauges (ụ-rô'-jēz)
urbanus (ûr'-bā-nus) belonging to the city, re-
fined, elegant,
urbicus (ûr'-bi-kus) belonging to the city.
Urceolaria* (ûr-sẽ-ôl-ā'-ri-ạ)
urceolate (ûr'-sē-ộ-lāt)
Urceolina* (ûr-sẽ-ol'-in-ạ, ûr-sẽ-ộ-lī'-nạ)
urceus (ûr'-se-us) a pitcher.
uredinia (ụ-rēd-i'-ni-ạ)
uredinous (ụ-rēd'-i-nus)
uredospore (ụ-rē'-dộ-spôr)
urens (ū'-renz) burning.
ureter (ū-rē'-têr)
urethra (ū-rē'-thrạ)
Urginea* (ûr-jin'-e-ạ)
Uria (ū'-ri-ạ)
Uroaëtus (ū-ro-ā'-ẽ-tus)
Urochroa (ū-rok'-rộ-ạ)

Urocichla (ū-rô-sik′-lạ)
Urocoptis (ū-rô-kop′-tis)
Urocyon (ū-ros′-i-on)
Urodela (ū-rô-dē′-lạ)
Urogale (ū-rog′-a-lē)
Uromastix (ū-rô-mas′-tiks)
Uromyces* (û-rom′-is-ēz)
Uromycladium* (û-rô-mī-klā′-di-um)
Uroplates (ū-rô-plā′-tēz)
Uropsilus (ū-rop′-si-lus)
Urosaurus (û-ros′-ôr-us)
Urospermum* (û-ros-pêr′-mum)
Ursinia* (ûr-sin′-i-ạ)
Urtica* (ûr-tī′-kạ)
Urubitinga (oō-roō-bi-tin′-gạ)
urubu (oō-roō-boō′)
Urvillea* (ûr-vil′-e-ạ)
usitatissimus (ū-si-ta-tis′-i-mus) most ordinary,
   very common.
Usnea (us′-ne-ạ)
Usofila (ū-sof′-il-ạ)
Ustilagnales (us-ti-lag-nā′-lēz)
Ustilago (us-ti-lā′-gō)
ustulatus (us-tu-lā′-tus) burned, scorched.
Uta (ū′-tạ)
Utricularia* (ū-trik-u-lā′-ri-ạ)
uvula (ū′-vu-lạ)
Uvularia* (ū-vû-lā′-ri-ạ)

V

Vaccinium* (vak-sin′-i-um, vak-sī′-ni-um)
vagans (vag′-anz) uncertain, wandering.

vagina (vaj-ī′-nạ)
vaginal (vaj′-i-nal, vaj-ī′-nal)
vaginalis (vaj-in-āl′-is)
vaginatus (vaj-i-nā′-tus) sheathed.
Vaginularia* (vaj-i-nul-ā′-ri-ạ)
vagrant (vā′-grant)
Valdesia* (val-dē′-shi-ạ, val-dē′-si-ạ)
Valeriana* (va-ler-i-ā′-nạ)
Valerianella* (va-ler-i-ạ-nel′-ạ)

Vampyrum <Fr. *vampire* =G. *vampyr*. Generic name of the blood-sucking
    bats. Pronounced: vam′-pi-rum, not vam-pī′-rum.

validus (val′-i-dus) strong, stout, vigorous.
Vallisneria* (val-is-nē′-ri-ạ)
Vallonia (val-ōn′-i-ạ)
Vampyrum (vam′-pi-rum)
Vanda* (van′-dạ)
Vanellus (van-el′-us)
vanessa (vạ-nes′-ạ)
Vanquelina* (van-kwe-lī′-nạ)
Varanus (var′-ạ-nus)
varicosus (var-i-kō′-sus) full of dilated veins.
variegatus (var-i-e-gā′-tus) of various colors, vari-
    ous, manifold.

varius (vā'-ri-us) diverse, changing, mottled.
Varonia* (vā-rō'-ni-ạ)
vas deferens (vas de'-fe-renz)
vasectomy (vas-ek'-tồ-mi)
velate (vē'-lāt)
velatus (vē-lā'-tus) furnished with a veil.
Velella (vē-lel'-ạ)
velifer (vē'-li-fêr) bearer of a veil.
veliger (vē'-lij-êr, vel'-ij-êr)
vellerosus (vel-er-ō'-sus) full of fleece.
Velozianum* (vel-ồ-zi-ā'-num)
velox (vē'-loks) swift-footed, quick.
velum (vē'-lum) a covering, a curtain.
velutinus (vel- û-tī'-nus) velvety, smooth.
vena cava (vē'-nạ-kā'-vạ)
venation (ven-ā'-shun)
venenatus (ven-ē-nā'-tus) poisonous.
Venerupis (ven-ệ-rū'-pis)
venetus (ven'-e-tus) sea-colored, bluish.
Venidium* (ven-id'-i-um)
Ventilago* (ven-til-ā'-gō)
venule (ven'-ūl)
venulosus (vē-nul-ō'-sus) full of small veins.
venustus (ven-us'-tus) lovely, pleasing, graceful,
    elegant.
Veratrum* (vē-rā'-trum)
Verbascum* (vêr-bas'-kum)
Verbena* (vêr-bē'-nạ)
Verbesina* (vêr-bes-ī'-nạ)
verecundus (ver-ē-kun'-dus) modest, shy.
Veretillum (ver-e-til'-um)
Vermes (vêr'-mēz)

vermiculatus (vêr-mik-ul-ā'-tus)

Vermivora (vêr-miv'-ôr-ạ)

vernalis (vêr-nā'-lis) of spring.

vernicosus (vêr-ni-kō'-sus) with surface appearing
   as if varnished.

Vernonia* (vêr-nō'-ni-ạ)

vernus (ver'-nus) of or belonging to spring.

Veronica* (vẹ-ron'-i-kạ, ver-on-ī'-kạ)

verrucosus (ver-ū-kō'-sus) full of warts.

versabilis (ver-sā'-bi-lis) changeable, movable.

versatilis (ver-sā'-til-is) able to be turned around,
   revolving, movable.

versicolor (ver-sik'-ul-or) of various colors.

vertagus (ver'-ta-gus) a gray-hound.

vertebra (vêr'-tẹ-brạ)

vertebral (vêr'-tẹ-bral)

verticil (ver'-ti-sil)

verticillatus (ver-ti-sil-ā'-tus) disposed in verticils,
   whorled.

vesica (vē-sī'-kạ, ves'-ik-ạ)

Vesicaria* (vē-sī-kā'-ri-ạ, ves-ik-ā'-ri-ạ)

vesicarius (vē-sī-kā'-ri-us) belonging to the bladder, curing pain in the bladder.

vespertine (ves'-pêr-tīn)

vespertinus (ves-pêr-tī'-nus) belonging to evening;
   also, western.

Vespidae (ves'-pi-dē)

vestibular (ves-tib'-ŭ-lâr)

vestigial (ves-tij'-i-al)

vestitus (ves'-tit-us) dressed, attired.

Vetiveria* (vet-i-vē'-ri-ạ)

vexillarius (vex-il-ā'-ri-us) like a flag.

vexillum (vek-sil'-um)

vial (vī'-al)

viaticus (vī-ā'-ti-kus) belonging to a road.

vibeks (vī'-beks) the mark of a blow, a stripe.

Viburnum* (vī-bûr'-num)

Vicia* (vish'-i-ạ, vis'-i-ạ)

vicinior (vis-in'-i-ôr)

Vidua (vid'-ŭ-ạ)

Viguiera* (vi-gwi-ē'-rạ)

Vilfa* (vil'-fạ)

villosus (vil-ō'-sus) hairy, rough, shaggy.

vimen (vī'-men) a switch, an osier.

viminalis (vim-i-nā'-lis) bearing or belonging to twigs for wickerwork.

Vinca* (ving'-kạ)

vinctus (ving'-tus) banded.

vinealis (vī-ne-āl'-is) of or belonging to vines.

vinnulus (vin'-ul-us) delightful, sweet.

Viola* (vī'-ô-lạ)

violaceus (vī-ô-lā'-se-us) violet-colored.

Viorna* (vī-ôr'-nạ)

Vipionidae (vip-i-on'-i-dē)

virens (vir'-enz) becoming green, verdant.

Vireo (vir'-e-ō)

virescens (vir-es'-senz) greenish, turning green, prospering.

virescent (vir-es'-ent)

virgatus (vir-gā'-tus) slender like a virga or rod.

viridis (vir'-i-dis) green; also, vigorous.

viridulus (vir-i'-du-lus) light green, somewhat green.

virosus (vir-ō'-sus) fond of men; also, full of slime, fetid, poisonous.
virulent (vir'-û-lent)
Viscacha (vis-kä'-chạ)
viscarius (vis-kā'-ri-us) bird-lime, slimy.
viscosus (vis-kō'-sus) sticky, viscous.
Viscum* (vis'-kum)
visnaga (vis-nä'-gạ)
vison (vī'-son)
vitality (vī-tal'-i-ti)
vitellin (vī-tel'-in)
vitelline (vī-tel'-ēn)
vitellus (vit-el'-us)
Vitex* (vī'-teks)
vitifolius (vī-ti-fol'-i-us, vī-ti-fō'-li-us) with vine-like leaves.
Vitis* (vī'-tis)
vitta (vīt'-ạ) a band.
vittatus (vit-ā'-tus) striped.
vivax (vī'-vax) long-lived, tenacious of life; also, vivacious, lively.
Viverra (viv-êr'-rạ, vī-ver'-ạ)
Viverricula (viv-er-ik'-ûl-ạ)
Vivipara (vī-vip'-à-rạ)
viviparous (vī-vip'-à-rus)
volador (vo'-la-dôr) a flier.
volans (vo'-lanz) flying.
volitans (vol'-i-tanz) flying.
volubilis (vol-ū'-bi-lis) twining, able to climb.
volucellus (vol-û-sel'-us) small-winged.
volvaceus (vol-vā'-se-us) covered by an external wrapper.

Volvox (vol'-voks)
Vombatus (vom'-bat-us)
vulgaris (vul-gā'-ris) usual, common, common-
place.
vulgatus (vul-gā'-tus) generally known, ordinary.
Vulpes (vul'-pēz)
vulpinus (vul-pī'-nus) of or belonging to a fox, fox-
like.

W

Wallabia (wäl-äb'-i-ạ)
Wallacei (wol-ā'-se-ī)
Weigela* (wī'-ge-lạ)
Welwitschia* (wel-wit'-chi-ạ)
Whipplea (whip'-lê-ạ)
Whitlavia* (whit-lā'-vi-ạ)
wislizeni (wis-liz-ē'-nī)
Wislizenia* (wis-li-zē'-ni-ạ)
Wissadula* (wis-ad'-du-lạ)
Wistaria* (wis-tā'-ri-ạ)
Wyethia* (wī-eth'-i-ạ, wī-ē'-thi-ạ)

X

Xanthisma* (zan-this'-mạ)
Xanthium* (zan'-thi-um)
Xanthocephalus (zan-thô-sef'-al-us)
Xanthocoma* (zan-thok'-ôm-ạ)
Xantholaema (zan-thô-lē'-mạ)
Xanthorrhiza* (zan-thô-rī'-zạ)
Xanthorrhoea* (zan-thô-rē'-ạ)
Xanthosoma* (zan-thô-sō'-mạ)
Xanthoxalis* (zan-thok'-sa-lis)

Xanthoxylum* (zan-thok′-si-lum)
Xantusia (zan-tū′-si-ạ)
Xema (zē′-mạ)
xenoecic (zen-ē′-sik)
Xenophonta* (zen-of-on′-tạ)
Xenopsilla (zen-op-si′-lạ)
Xeranthemum* (zē-ran′-the-mum)
xeric (zē′-rik)
xerochasy (zē-rô-kā′-si)
Xerophyllum* (zē-rô-fil′-um)
Xerophyta* (zē-rof′-it-ạ)
xerophyte (zē′-rôf-īt)
xerophyton (zē-rof-ī′-ton)
xeropoium (zē-rôp-ō′-i-um)
xerosere (zē′-ros-ēr)
Xestobium (zes-tō′-bi-um)
Ximenia* (zī-mē′-ni-ạ)
Xiphidium* (zif-id′-i-um)
xiphihumeralis (zif-i-hū-mer-ā′-lis)
Xiphosura (zif-ôs-ū′-rạ)
Xyelidae (zī-el′-i-dē)
Xylaria* (zī-lā′-ri-ạ)
Xyleborus (zī-leb′-ôr-us)
xylesthia (zī-les′-thi-ạ)
Xyleutes (zī-lū′-tēz)
Xylia (zī′-li-ạ, zil′-i-ạ)
Xylobium* (zī-lob′-i-um, zil-ob′-i-um)
Xylocopa (zī-lok′-ô-pạ)
Xylocopidae (zī-lô-kôp′-i-dē)
Xylophagus (zī-lof′-ag-us)
Xylophylla* (zī-lô-fi′-lạ)
Xyrauchen (zī′-rô-kēn)

316

Xylocopa <Gr. *xylos*, wood+*tomŏ*, to cut. Generic name of the wood cutting bees. Pronounced: zī-lok'-ŏ-pạ, not zī-lō-kŏ'-pạ.

Xyris* (zī'-ris, zir'-is)
Xysticus (zis'-ti-kus)

Y

yolk (yōk, yōlk)
Yponomeutidae (ip-ŏ-nŏ-mūt'-i-dē)
Yucca (yu'-kạ)

Z

Zaglossus (zag-los'-us, zag-lō'-sus)
Zaitha (zā'-thạ)
Zalophus (zal'-ŏ-fus)
Zamenis (zam'-e-nis)
Zamia (zā'-mi-ạ)
Zanclus (zang'-klus)
Zanonia* (zā-nō'-ni-ạ)
Zapus (zā'-pus)
Zea* (zē'-ạ)
Zelotes (ze-lō'-tēz)
Zenobia* (zen-ō'-bi-ạ)
Zephyranthes* (zef-i-ran'-thēz)
zerda (zêr'-dạ)
Zeus (zē'-us)
Zeuzera (zŭ-zē'-rạ)

Zeuzeridae (zū-zer′-i-dē)
Zibethailurus (zi-beth-āl-ū′-rus)
zibethicus (zi-beth′-i-kus)
Zingiber* (zin′-ji-bêr)
Ziphius (zif′-i-us)
Zizania* (zī-zā′-ni-ạ)
Ziziphus* (ziz′-i-fus)
zoarium (zō-ā′-ri-um)
Zodion (zō′-di-on)
Zoea (zō-ē′-ạ)
zoecium (zō-ē′-shi-um)
zoehemera (zō-ê-hem′-er-ạ)
Zonotrichia (zō-nȯ-trik′-i-ạ)
Zonurus (zō-nū′-rus)
zoology (zō-ol′-ȯj-i)
Zoomastigina (zō-ȯ-mas-ti-jī′-nạ)
zoophilous (zō-of′-i-lus)
zootomy (zō-ot′-ȯ-mi)
Zoraptera (zôr-ap′-te-rạ)
Zostera* (zos-tē′-rạ)
Zoysia* (zoy′-si-ạ)
Zygadenus (zī-gad′-e-nus, zig-ad-ē′-nus)
Zygogeomys (zī-gȯ-jē′-ȯ-mis, zig-ȯ-jē-o′-mis)
zygomorphic (zī-gȯ-môr′-fik, zig-ȯ-môr′-fic)
zygospore (zī′-gȯ-spôr, zig′-ȯ-spôr)
zygote (zī′-gōt)
zymolysis (zī-mol′-is-is)